Dyes in History and Archaeology 35/36

Editor's note

Dyes in History and Archaeology is a refereed journal; the papers included are largely drawn from those presented at the annual meetings of *Dyes in History and Archaeology* and are accepted for publication, after peer review, on the assumption that a paper submitted does not contain essentially the same material as in a previous publication, and is not under consideration for publication elsewhere.

The journal was formerly published by Textile Research Associates, York (issue 7, 1989 to issue 15, 1997).

Dyes in History and Archaeology 35/36

Including papers presented at the 35th Meeting held at Pisa, Italy, 6–8 October 2016 and the 36th Meeting at Hampton Court, UK, 26–28 October 2017

Edited by Jo Kirby
Editor-in-Chief, Dyes in History and Archaeology

with the assistance of
Christopher J. Cooksey, *Independent researcher, Watford*
Vanessa Habib, *Independent researcher, Edinburgh*
Catherine Higgitt, *Scientific Department, The National Gallery, London*
Richard Laursen, *Professor Emeritus, Department of Chemistry, Boston University*
Anita Quye, *Centre for Textile Conservation and Technical Art History, University of Glasgow*
Terry Schaeffer, *Los Angeles County Museum of Art*
and Maarten van Bommel, *Professor, Department of Conservation and Restoration, University of Amsterdam*

Archetype Publications

First published 2021 by Archetype Publications Ltd

Archetype Publications Ltd
c/o International Academic Projects
1 Birdcage Walk
London SW1H 9JJ
www.archetype.co.uk

ISBN: 978-1-909492-81-3

British Library Cataloguing in Publication Data
A catalogue record for this book is available from the British Library.

Printed on acid-free paper

MIX
Paper from responsible sources
FSC® C020438

Typeset by Marcus Nichols at PDQ Digital Media Solutions Ltd, Bungay
Printed in the UK by Hobbs the Printers Ltd, Totton, Hampshire

Contents

Editorial vii

Cornelis Drebbel's Scarlet 1
Matthijs de Keijzer, Art Néss Proaño Gaibor and Muriel Geldof

Fake Lakes? The Use of Starch and other Adsorbent Substrates for Synthetic Dyes in Paint Manufacture 15
Eva Eis

Treading on Lichen: The Use of Cudbear by a Victorian Carpet Manufacturer 25
Terry T. Schaeffer and Laura Maccarelli

A Non-Destructive Approach to Dye Analysis for Typical 17th–19th-Century East Asian Textiles 39
Yoshiko Sasaki and Ken Sasaki

Non-Destructive Spectroscopic Approaches for the Analysis of Dyes in Japanese Textiles: Analyses of an Early Edo Era (17th Century) Kimono Fragment 50
Yoshiko Sasaki and Ken Sasaki

The Tradition of Using Natural Dyes in Latvia 62
Anete Karlsone

Revealing the Mysteries of the Madder Bath 70
Vincent Daniels

Analysis of Red Yarns in Textiles from the Huaca Malena Archaeological Site in Peru 78
Laura Maccarelli, Terry T. Schaeffer, Elizabeth Burr and Ioanna Kakoulli

Ancient Dyes with Medical and Pharmacological Properties 89
M. Julia Martínez García, M. Luisa Vázquez de Ágredos Pascual and Rita Pagiotti

Yellow Dyes of Historical Importance IV: Sourcing and Consumption of Yellow Dyes by the Dyeing Company, Francesco di Giuliano Salviati e Comp., tintori d'Arte Maggiore, in Florence (1483–1498) 108
Dominique Cardon and Ingrid Houssaye Michienzi

Natural Yellow Dye Sources in Persian Carpets: A Review 121
Samaneh Sharif, Paula Nabais and Maria João Melo

Editorial

This issue of the journal contains papers presented at the 35th meeting of *Dyes in History and Archaeology*, held at the University of Pisa in 2016, and the 36th meeting, held at the Royal School of Needlework, Hampton Court Palace, in 2017, together with some from earlier meetings that were submitted more recently.

The range of topics covered is very broad, reflecting the varied interests of the programmes of lectures and of those attending. While the chemistry and identification of the dyes always play a significant part, the principal point of the analysis described is to give information on the artefact being studied, its history and the technology of its production. The discussion of a mechanism of dyeing, such as the madder bath described at Hampton Court, or of a method of analysis, such as the use of a particular reagent to derivatise the samples for analysis by gas chromatography–mass spectrometry described in Pisa, is still valuable, however, as it is likely to have a wider relevance.

In recent years attention has been devoted to the synthetic dyes of the late 19th and early 20th century, both as textile dyes and in pigments. In Pisa, for example, dyestuff patents, purple aniline dyes, the adsorbent substrates used for lake pigments and the fading of the lake pigment prepared from eosin were all subjects for discussion, while at Hampton Court, delegates learned about the interaction between early synthetic dyes and the paper used in sample books and the degradation of xanthene dyes, which include eosin. Natural dyes and the artefacts coloured with them were the subject of the majority of the papers at both meetings, however. The subjects covered during the Pisa meeting ranged from dyes traded by the late 15th-century Florentine dye company Francesco di Giuliano Salviati e Comp., tintori d'Arte Maggiore and red dyes in Peruvian archaeological textiles to lichen purple dyes and the deterioration of safflower dyeings. The topics covered at Hampton Court were equally varied, including an 8th-century Chinese embroidery, the study of 17th-century textiles found in a shipwreck in the Wadden Sea off the north-west coast of the Netherlands and Denmark and the amazing scarlet cochineal colour discovered by Cornelis Drebbel in the 17th century. A wonderfully varied display of posters was also presented at the two meetings and the papers published here reflect the variety of subjects covered.

As part of the meetings, visits were arranged to sites of interest. Delegates attending the Pisa meeting travelled to the city of Prato to visit the Duomo (Cathedral of Saint Stephen) and the Museo di Tessuto, which exhibits historical and contemporary textile production. In Pisa itself, however, delegates were able to visit the home of the Salviati Archive in the Scuola Normale Superiore and it is appropriate that one of the papers published in this book describes the activities of the Salviati company trading in sources of yellow dyes in the 1480s and 90s. As well as seeing something of the work of the Royal School of Needlework and the architecture of Hampton Court Palace itself, those who attended the 36th meeting could also visit the Dennis Severs house in Spitalfields, East London, an 18th-century house which has been decorated and furnished to illustrate aspects of the life of a fictional family of Huguenot weavers in a realistic way, and the Warner Textile Archive in Braintree.

On behalf of all those who attended the meetings I would like to thank the organisers, Maria Perla Colombini, Ilaria Degano and their colleagues at the University of Pisa, and Susan Kay-Williams and her colleagues at the Royal School of Needlework, Hampton Court, and the members of the selection committees for the technical content of the meetings, Dominique Cardon, Ilaria Degano, Regina Hofmann-de Keijzer, Jo Kirby, Richard Laursen, Maria João Melo, Irina Petroviciu, Erika Ribechini, Maarten van Bommel and André Verhecken.

My thanks are also due to my co-editors, Chris Cooksey, Vanessa Habib, Catherine Higgitt, Richard Laursen, Anita Quye and Terry Schaeffer, and in particular to Maarten van Bommel who has coordinated the whole process of contacting referees, authors and editors; Without their assistance the production of this volume would have been much more difficult and it would have been impossible without the help of our hosts for the two conferences, the University of Pisa and the Royal School of Needlework. Access to documentary sources, to artefacts and to illustrative material is of fundamental importance to successful research and to a beautiful publication that brings it to life. The people who have helped in this respect have been acknowledged in the individual papers, but I would like to thank some of them here as well: Maddalena Taglioli, curator of the Salviati Archive and Head Archivist, Centro Archivistico, Scuola Normale Superiore, Pisa; Prof. Dr Elisabeth Tietmeyer and Salwa-Victoria Joram, Museum Europäischer Kulturen, Staatliche Museen zu Berlin; and P.J. Chatzidakis, Director of the Archaeological Museum of Vathy, Samos, Greece. Last, but not least, I am most grateful, as always, to Jim Black and the team at Archetype Publications.

Note. All plant names have been checked using the International Plant Names Index IPNI (2020). International Plant Names Index. Published on the Internet https://www.ipni.org, The Royal Botanic Gardens, Kew, Harvard University Herbaria & Libraries and Australian National Botanic Gardens [accessed most recently 26 November 2020]. Other sources were accessed by way of the Royal Botanic Gardens, Kew, website. Similar checks have been made on the names of animal species, fungi and lichens where these occur.

Cornelis Drebbel's Scarlet

Matthijs de Keijzer,[1]* Art Néss Proaño Gaibor[1] and Muriel Geldof[1]

Abstract At the beginning of the 17th century, the Dutchman Cornelis Jacobszoon Drebbel brought about a great improvement in the dyeing of cloth from a chemical and technological point of view. He introduced the element tin as a mordant in cochineal dyeing, creating an intense scarlet colour. However, there are doubts about this invention and the year of its introduction. It is remarkable that Drebbel's Scarlet as a dye is hardly ever detected in textile objects by chemical analysis. Recipes for tin-containing pigments are known from the mid-18th century onwards, and in the 19th century artists such as Vincent van Gogh and his contemporaries used tin-containing lakes of cochineal. At the end of the 19th century, Drebbel's Scarlet was gradually replaced by synthetic organic dyes, especially the Ponceaus (2-naphthol-disulfonic acids).

Introduction

Since the 17th century, Cornelis Jacobszoon Drebbel has either been maligned or highly praised. In the 1746 *Kronyck van Alckmaar*, the chronicler Cornelis van der Woude named him '*een uytgenomen en boven maten uytmuntende vestandigh Philosooph: wiens hooghe vernuft, ende scherpsinnighe begryp (in zynen tijdt) in de gansche Werelt geen gelyck en hadde inde Philosophie*' ('an excellent wise philosopher, whose high ingenuity and understanding (in his days) had no equal in the world of philosophy'). At the beginning of the 20th century, the physicist and historian of science Henri Adrien Naber (1867–1944) described Drebbel, in various publications, in an exaggerated and often unscientific way as one of the most original and greatest minds of his time. In 1922, the chemist and historian of science Frans Maurits Jaeger (1877–1945) reacted fiercely against Naber's uncritical adoration: 'not only did Drebbel not enrich science with new ideas, but he was not even a serious scientist. He was no more than an adventurer and a pauper, who tried his luck all over Europe without finding it anywhere.' Ten years later, in his thesis of 1932, the physicist Gerrit Tierie was less judgemental: he made no mention of Drebbel as a great scientist but as undoubtedly an ingenious technologist.[1] Some of Drebbel's inventions were entertaining, while others were technologically interesting or useful, but one technological discovery attributed to him has been of great and lasting significance: he found that cochineal gave a scarlet dye if a tin mordant was used when dyeing, a discovery which was to prove of considerable importance for both textile dyeing and pigment making in the centuries to follow.

Cornelis Jacobszoon Drebbel: his life and technological inventions

Drebbel was born in the Netherlands town of Alkmaar in 1572. After an elementary education, including some years at the Latin School, around

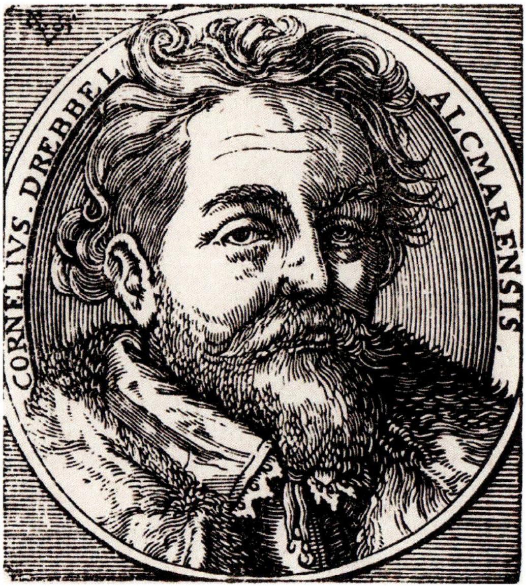

Figure 1 Christoffel van Sichem the Younger, *Cornelis Drebbel*, 1631, engraving, Collection Alkmaar Regional Archives/PR 1001628.

1590 he moved to Haarlem to live with the engraver Hendrick Goltzius (1558–1617) as his apprentice and be initiated into the art of alchemy. In 1595 Drebbel married Goltzius's younger sister, Sophia Jans, and settled in Alkmaar as a painter, engraver and cartographer (Fig. 1).

As an empirical researcher and innovator, Drebbel was granted two patents on 21 June 1598, the first covering a water supply system to lead large quantities of fresh water through lead pipes and to raise it, like a fountain, to heights of 30, 40 or 50 feet; the second for a perpetual motion clock. In 1601, he constructed a fountain outside the Noorderpoort (North Gate) in Middelburg, and on 16 February 1602 he received a further patent for his design of a chimney with a good draught. These three patents, the only ones Drebbel ever obtained, were granted by the States General of the Netherlands.[2]

After the death of the Tudor queen, Elizabeth I of England (1533–1603), she was succeeded by the Stuart king, James I (1566–1625). To restrict imports and to stimulate craftsmanship and local business, the king promoted the manufacture of silk, glass, tapestry, dyed cloth and other luxury goods by inviting skilled artisans from the Low Countries, France and Italy to come to England, including Cornelis Drebbel, one of the best-known inventors in Europe at the beginning of the 17th century.

In late 1604 or early 1605, Drebbel moved to England to join the court of James I and was soon taken into the service of the king's 10-year-old son, Henry Frederick Stuart, Prince of Wales (1594–1612). At Eltham Palace in Greenwich, south-east London, Drebbel astonished the court with his inventions including: a *perpetuum mobile* (perpetual motion machine); optical instruments; a clavichord which was played by the rays of the sun; and incubators to hatch duck and chicken eggs all year round, even in midwinter.[3] In 1607 Drebbel also established a dye-works in Stratford-at-Bow on the river Lea, 14 miles from London.

Drebbel's reputation soon spread to the courts of Europe: Lewis Frederick, Duke of Wurtemburg (1582–1628), visited him at work at Eltham Palace in May 1610, and in October of that year Drebbel accepted the invitation of the German emperor, Rudolf II (1552–1612), who was interested in alchemy, astrology and art, to come to his court in Prague. In April 1611, however, Rudolf was deposed by his younger brother Matthias (1557–1619) and died in January 1612. Drebbel lost his patron and was imprisoned for one year. Returning to London in February 1613, he discovered that Prince Henry had died in November 1612: losing his two main protectors within the period of a year resulted in financial difficulties for Drebbel.[4] An expert in lens grinding, he devoted several years to optics and the manufacture of microscopes, and in 1621 achieved widespread fame for his invention of a microscope with two convex lenses.

In a change of direction in 1620 he constructed a navigable submarine that was demonstrated to the king and thousands of Londoners: it could travel submerged in the Thames from Westminster to Greenwich for three hours at a depth of 15 ft.[5] From 1626 to 1628, Drebbel was employed by the British Navy to manufacture explosives such as the water petards and water mines used to relieve the Huguenot city of La Rochelle. The city was taken by the French troops in 1628 resulting in Drebbel's dismissal. According to an account in Rawlinson Manuscript B. 158 (Bodleian Library, Oxford), towards the end of his life Drebbel lived in near

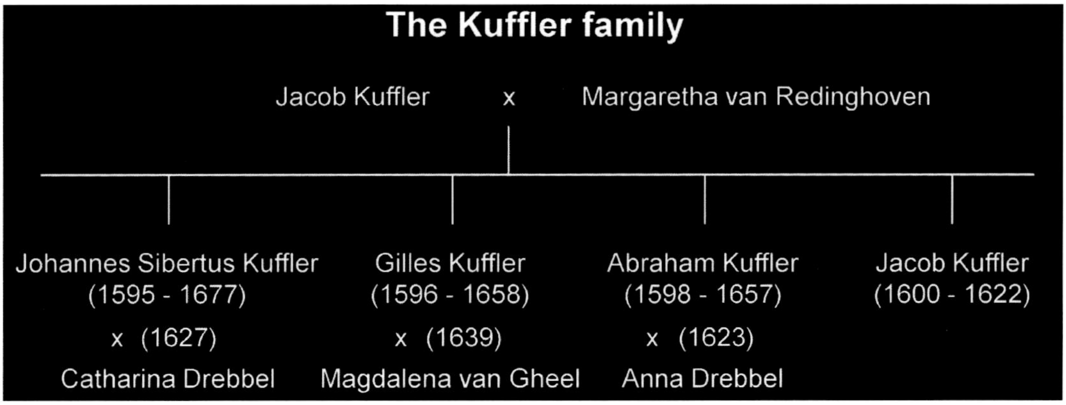

Figure 2 The Kuffler family and their connections with the Drebbels. (Photo © Matthijs de Keijzer)

poverty, earning his living as a brewer and innkeeper running an alehouse below London Bridge:

> He was very poore, and in his later time kept an Ale-house below the [London] bridge. He had an invention of going under water which he used so advantageously, that many persons were persuaded that he was some strange Monstar, and that means drew many to see him and drink of his ale.[6]

Drebbel died in the Parish of Trinity, Minories, in London in 1633.

An earlier scarlet dye

Before American cochineal was imported into Europe in the 16th century, the dye insects kermes and the so-called Polish and Armenian 'cochineals' were used for dyeing red. At the beginning of the 17th century, cochineal was employed with a mordant of alum (aluminium potassium sulfate) and cream of tartar (potassium hydrogen tartrate) to yield crimson shades. A brilliant and intense scarlet colour on wool and silk, called Venetian Scarlet, could be obtained using kermes mordanted with alum. Drebbel, known as the inventor of a scarlet obtained by mordanting cochineal with tin, contributed a great technological improvement in the dyeing of cloth. The process basically consisted of dissolving granulated tin in nitric acid (*aqua fortis*) to obtain the mordanting solution. Drebbel's Scarlet replaced the kermes and alum-based Venetian Scarlet as the most common, most desirable, bright red colour. The history and development of Drebbel's Scarlet is closely linked with four brothers, the Kufflers, who played an important role in his life.

The four Kuffler brothers and their dye-works

The brothers were the sons of Jacob Kuffler (d.1607) and his wife Margaretha van Redinghoven, who had moved from Holland to Cologne for religious reasons. Around 1620, the brothers Abraham (1598–1657) and Jacob Kuffler (1600–1622) came to England in the hope of gaining the favour of James I by offering him a remarkable book but were not successful. Seeking help from Drebbel, some time later they became business associates in Drebbel's dye-works. In 1622, Jacob was sent to France and Italy to source a market for Drebbel's microscopes and lenses, but he died of plague at Rome in November of that year.

Following his marriage to Drebbel's daughter Anna in 1623, Abraham sent for his two other brothers, Johannes Sibertus (1595–1677) and Gilles or Aegidius (1596–1658). In 1627, Johannes married Drebbel's second daughter Catharina (Fig. 2). The brothers went into business with their father-in-law, acting as agents and promoters of Drebbel's inventions both in England and on the Continent. Gilles, who came from Cologne, had been apprenticed to an uncle, Abraham Kuffler, a cloth merchant in Amsterdam. After working for some time in England he then settled permanently in the Netherlands.[7] After Drebbel's death in 1633, the three Kuffler brothers converted the dye-works at Stratford-at-Bow into a company in 1635 and renamed Drebbel's innovation in colour 'Kufflerianus to dye wool and silk scarlet'. The recipe remained a well-guarded family secret and in 1638 Abraham became a naturalised Englishman.[8]

Probably for political reasons due to the outbreak of the English Civil War, Johannes Sibertus and Abraham, together with their whole ménage, left England in June 1642 and settled in the Netherlands, initially in The Hague, taking their scarlet process with them.[9] Here too the Kuffler brothers began to devote themselves to dyeing, first in Katwijk on the Old Rhine in the neighbourhood of Leiden and some time later in Arnhem. A deed drawn up in Leiden shows that in 1647 a certain Jacob Kuffler with a certain Gys as his partner, set up a dye-works at Katwijk dyeing scarlet cloth. Katwijk may have been chosen over Leiden because the new method demanded the use of very clear water.[10]

In addition, the Kuffler brothers, together with their cousins Johannes Kuffler (1608–1661) (son of Abraham Kuffler, the abovementioned Amsterdam cloth merchant) and Abraham Velthuysen, established a dye-works in their joint names at Hulckensteyn, near Arnhem. Velthuysen, a merchant from Huil, was appointed its director. From the start the company was plagued by all kinds of difficulties and violent quarrels. Velthuysen was held responsible for the damage caused by unacceptable results ('*twe ende seventich duysent guld.*, 72.000 guilders'), but he claimed that these were due to the incomplete instructions he received from Johannes Sibertus. From documents in the archives in Arnhem it is clear that Sibertus knew how to dye a good scarlet. In 1649, Johannes Kuffler, against the wishes of his uncle, Johannes Sibertus, resigned and started his own business at Zeist. After a trial conducted by the Kufflers against their cousin Velthuysen in 1652 the case was finally settled amicably in 1654.[11]

Gilles married Magdalena van Gheel at Delft in 1639, remaining in the Netherlands while his brothers later returned to England. According to another Leiden deed he had settled in Amsterdam between 1649 and 1651, probably at the house of his uncle Abraham Kuffler, but it appears that he was not well versed in the art of scarlet dyeing.[12] In 1655, Abraham returned to England and the dye company was again based in Stratford-at-Bow. Following Abraham's death in 1657, the business was taken over by Johannes Sibertus (who in subsequent years was appointed Physician-in-Ordinary to the Duke of York) until he died in London at the age of 82 in 1677.[13] It was Johannes Sibertus in particular who, after Drebbel's death, did most to commercialise his inventions.

Drebbel's Scarlet as a dye

Discovery of the method and its date

Unfortunately, Drebbel left only a few writings, such as *Een Kort Tractaet van de Natuere der Elementen*, published in 1604, and *De Quinta Essentia Tractatus*, in 1621, without describing his Scarlet, so most data on the dye are obtained from texts by Drebbel's contemporaries and from secondary texts, six of which were written much later after his death: the writings of Robert Boyle (1627–1691), John Evelyn (1620–1705), Fridericus Hoffmann (1660–1742), Richard Rawlinson (1690–1755) and Johann Beckmann (1739–1811). In these texts Drebbel is mentioned as the inventor of Scarlet dye, also named 'bodied scarlet' or 'died scarlet'.[14]

Beckmann acquired the history of Drebbel's Scarlet from one of the most learned of German dyers, M. Kuhlenkamp at Bremen. The story is as follows:

> Drebbel having placed in his window an extract of cochineal, made with boiling water, for the purpose of filling a thermometer, some *aqua regia* (a mixture of three parts concentrated hydrochloric acid and one part concentrated nitric acid) dropped into it from a phial by accident, which stood above it, and converted the purple dye into a most beautiful bright scarlet. After some conjectures and experiments, he discovered that the tin by which the window-frame was divided into squares had been dissolved by the *aqua regia*, and was the cause of the change. He (Drebbel)

> communicated this observation to Kuffelar, an ingenious dyer at Leyden, who was afterwards his son-in-law. The latter brought the discovery to perfection and employed it some years alone in his dye-house, which, gave rise to the name of Kuffelar's colour.[15]

The Kuffler family records reveal that the Stratford-at-Bow dye-works established by Drebbel in 1607 did not flourish. That it really was Drebbel who was responsible for the scarlet invention is confirmed by his contemporary, the French savant Nicolas-Claude Fabri de Peiresc (1580–1637). Although he never met Drebbel personally, he obtained his information from the Kuffler brothers, who paid him several visits in Paris between 1624 and *c.*1627. At the beginning of September 1624, Abraham and Gilles travelled to Paris for a meeting with Peiresc during which the Kuffler brothers told him about the microscopes, lenses and the submarine; Drebbel's Scarlet was not mentioned. Some years later, in 1627, Johannes Sibertus met Peiresc and showed him specimens of dyed scarlet. Peiresc writes: 'Drebbel has shown them a way of making the scarlet dye, and this is more beautiful and at a better price than the colours now usually made. Kuffler showed me specimens of it which were remarkable fine.'[16]

However, there are doubts about the scarlet invention and the year of discovery. The literature cites different dates: 'in the beginning of the 17th century'; 'in the early days of the 17th century'; 'the period 1606–1607'; 'in 1607'; '*c.*1610'; 'around 1625; 'the period 1625–1627'; 'as early as 1627' – even the year 1630 is mentioned. According to Beckmann, the discovery took place in 1630, while Jaeger is of the opinion that the date should be 1607. The greatest obstacle is that Abraham, and Johannes Sibertus in particular, tended to encourage mystery and exaggerated Drebbel's inventions, including his Scarlet, to the extent that it is extremely difficult to determine the facts. The mystery is further enhanced by Peiresc, who wrote that Johannes Sibertus possessed Drebbel's laboratory book with his inventions.

From these textual references it is possible to conclude that the invention of Drebbel's Scarlet must have taken place between 1607, when his dye-works was founded, and 1627, when Johannes Sibertus met Peiresc and showed him specimens of dyed scarlet. The fact that Drebbel's Scarlet is not described in *De Quinta Essentia Tractatus*, published in 1621, and was not mentioned in 1624, when Abraham and Gilles met Peiresc for the first time in Paris, makes it probable that this dye method was developed between 1624 and 1627. As already mentioned, according to the Kuffler family records, the dye-works set up in 1607 did not prosper but Jaeger is convinced that the date of invention should be 1607. If this were the case, such an invention would be a tremendous boost to the dye-works, which would then have become successful and highly profitable. And why wait so long to reveal this technique? However, it cannot be excluded that Drebbel had either invented or improved the technique before 1624–27.

Tin salts and pewter vessels

There is another issue: did Drebbel really apply a tin salt as a mordant or use a pewter vessel – or both? Pewter is an alloy primarily consisting of tin with a small amount of copper and other metals (lead in early pewter, more recently antimony). In 1662, William Petty (1623–1687), the son of a Hampshire cloth merchant, who collected information on textiles and methods for dyeing, gave a lecture at the Royal Society of London entitled 'An Apparatus to the History of the Common Practises of Dying', from which it is evident that pewter was used as a mordant in the dye-works. On pewter, Petty reports that 'They also use Pewter for Bow-dye, Scarlet, *viz.* they dissolve Bars of Pewter in the *Aqua fortis* they use; and make also their Dying-kettles or Furnace of this Metall.' Concerning cochineal he remarks: 'This also is used with Bran-liquor in Pewter Furnaces, and with Aqua-fortis, in order to the Scarlet-dye [*sic*].' Referring to the vessels or tubs, he observes: 'In some cases the matter of the Vessel wherein the Liquors are heated, and the Tinctures prepared, must be regarded; as the Kettles must be Pewter for Bow-dye.'[17]

In 1667 Robert Boyle relates that: 'A most famous master in this art (possibly Johannes Sibertus Kuffler) assured me, that neither he, nor others can strike the lovely colour, called bow-dye, unless their materials be boiled in "vessels of a particular metal"' and continues 'Thus in dying scarlet, ... that the vessels are to be made of, or lined with tin; you would never be able to bring the tincture of cochineal to give a perfect scarlet.'[18] The German alchemist Johannes Kunckel (1630–1703) writes in his 1679 *Ars vitraria experimentalis*: 'It is possible to obtain this colour in another way namely with pewter and strong water and in pewter vats producing a much greater degree of brightness.'[19] In 1702, Fridericus Hoffmann records in his *Opera Omnia*: 'For the discovery of scarlet dye is ascribed to Drebbel, the famous chemist from Holland, who made a solution of cochineal in aqua fortis, and in order to prevent this from eating away the fibre of the cloth, he added water and pewter filings to mitigate the acidity.'[20] According to *Le teinturier parfait*, published in 1708, the method to dye cloth with Dutch scarlet was the same as the Kufflers used in England.[21]

Dyers commented that pewter had an influence on the scarlet dye. By April 1648, Jan Hennebo of Leiden had already noticed that scarlet dyed in pewter vessels was much brighter and more beautiful than scarlet dyed in brass vessels.[22] The 1709 volume *Der Curieuse und volkommene Kunst Faerber* contains the sentence, 'Any one wishing to dye scarlet must above all make a special point of obtaining a heavy kettle made of pure unadulterated pewter (large or small, according to the amount he wishes to dye) and to be sure to keep this clean always.'[23]

In the 1666–1690 Cambridge manuscript compiled by one of Drebbel's grandsons, Augustus Kuffler (born in The Hague in 1644, the son of Johannes Sibertus and Drebbel's daughter Catharina), provides information on the scarlet dye. Drebbel's name is not mentioned, the colouring matter is cochineal and none of the recipes mentions the use of tin salts when dyeing scarlet: only in a single passage does the word 'pewter' appear, mentioned in relation to a pottinger (a tin can). On the other hand, the manuscript contains a recipe entitled 'To make the Secrett' that mentions the word 'Emitt'. Tierie has suggested that this word may be used to conceal the use of some form of tin salts. [24] However, the engineer and author Lawrence Ernest Harris (1892–1959) drew the conclusion that although this method had not been invented by Drebbel, the use of tin salts was still a closely guarded secret by the Kuffler dye-works. At the time of the compilation it is unlikely that Augustus Kuffler would have disclosed the method to all and sundry.[25]

According to the last three references, it is less certain that Drebbel used a tin salt as a mordant: some think that Drebbel probably discovered this dyeing method serendipitously while others even believe that he is not the inventor but rather the innovator of the process of using tin salts as mordants for fixing the dye on the fabric, and for giving the colour a much greater brilliance.

Scarlet dyeing in practice

Regardless of the doubts, the Kufflers brought the scarlet method to the Netherlands and by the 1650s the process had disseminated across Europe. The dye became known under various names such as Drebbel's Scarlet, Dutch Scarlet, Holland Scarlet, Bow dye, Scarlet Bow dye, Bow-dyed Scarlet, bodied scarlet, died scarlet, Hunting Scarlet, Fire Scarlet, Kuffler's Red, Kuffelaars and colour Kufflerianus. It was considered to be the finest for scarlet dyeing with a brilliant orange-red colour.

In France, after the Gobelin family in Paris had given up dyeing in what had been their dye-works, their place was taken by two families: the Comans and the de la Planches (Flemish tapestry makers). In 1629, the families parted company but the Comans continued their activities at the dye-works until 1654. Then in 1655, the Dutchman Jean Gluck (Glucq or Gluq) from Amsterdam introduced the Drebbel's scarlet 'à la hollandaise' method to colour wool and cloths from Amsterdam into the Maisons des Gobelins.[26] He must have been initiated into the secrets by Gilles Kuffler, then living in Amsterdam. Around 1660, Jean Gluck acquired the dye-works and integrated the Dutch method. According to Tierie, Gluck was in contact with the Kufflers in Amsterdam for quite some time and that it is probable that the engraver Nicolaes Glock (or Kloeck) who worked with Drebbel had a family connection to the cloth maker Kloeck (or Gluck) from Amsterdam. This explains why scarlet dyeing was transferred to the Manufactures des Gobelins via the Kufflers.[27]

Jean-Baptiste Colbert (1619–1683), Louis XIV's minister of finance and a man who had come from a family of cloth makers and merchants in Rheims,

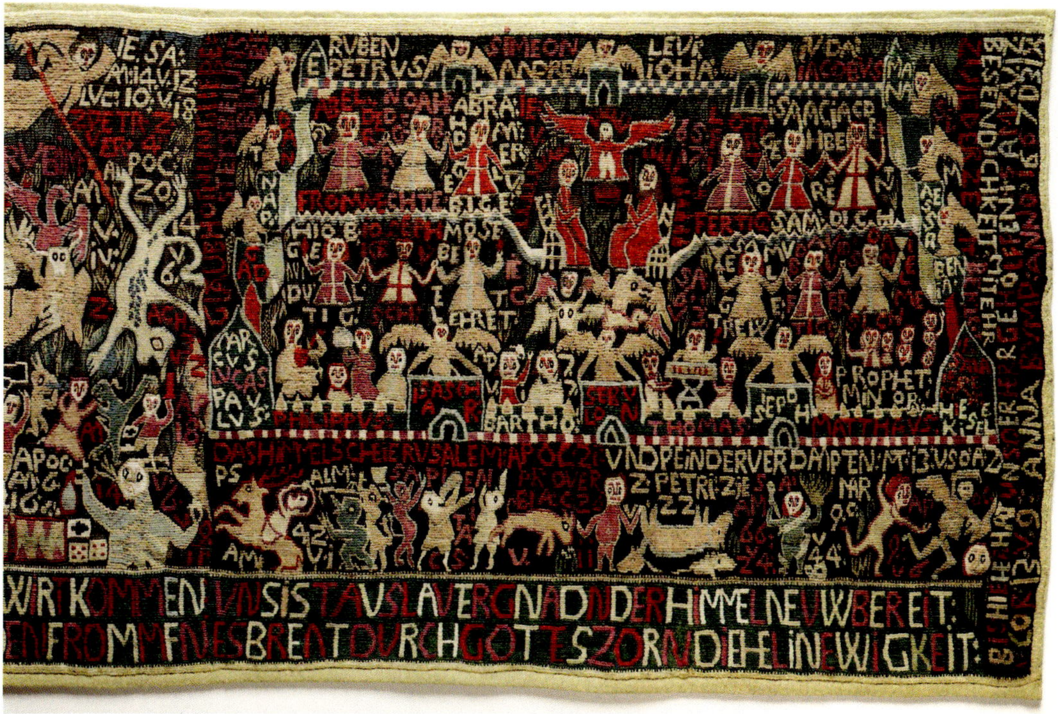

Figure 3 Anna Bump, *Reformation Tapestry, Commemorating the 150th Anniversary of the Reformation*, detail of Field V (extreme right): *The Heavenly Jerusalem*, 1667. Made in Hennstedt, Northern Germany. Warp, hemp; weft, differently coloured wools, silk and unbleached linen; 56 × 350 cm (whole tapestry). Museum of European Cultures, Berlin, Ident. Nr. D (17 J 18) 745/1971. (Photo © Staatliche Museen zu Berlin, Museum Europäischer Kulturen/Ute Franz-Scarciglia)

and an identified M. d'Albo have both been credited with the authorship of an anonymous dye book entitled *L'instruction générale pour la teinture de laines et manufactures de laine de toutes couleurs*, published in Paris in 1671. Seven types of *bon teints* for red are mentioned: the old *procédé de l'écarlate de France* (French Red Scarlet or Gobelins Red Scarlet or *graine scarlet*) referring to kermes; Cramoisy Red (crimson); Madder red; Half-graine; Half-Cramoisy; *Nacarat de bourre* (flock or shred red, using dye derived from cloth clippings) and *rouge écarlate* (Cochineal scarlet, 'as made in Holland') prepared with tin as a mordant.[28] Furthermore, Jean Gluck obtained letters patent from Louis XIV in 1677 permitting him to establish works at Lyons for the bleaching and dyeing of scarlet cloth and linen wares according to the Dutch method for 20 years.[29]

In the 18th century, cochineal red also found its way to the English colonies in North America. The dye books of Elijah Bemiss, J. and R. Bronson and William Partridge, written at the beginning of the 19th century, confirm that the method to prepare and dye with Scarlet Red was well known.[30] The 1823 dye book of William Partridge mentions five recipes for dyeing scarlet on wool and silk. He describes that it was necessary for the dyers to prepare their own soluble tin mordant, making it difficult to determine the exact amount of tin chloride they were using. To make the tin liquor, the only ingredient of a regular quality is the metal tin, imported as stamped blocks from Cornwall, by far the main producer in Europe at the time. John Edmonds (1931–2009) and a team of dyers performed dyeing experiments following Partridge's recipes. American cochineal (*Dactylopius coccus* Costa) dye was confirmed as being present using ultra high-performance liquid chromatography (UHPLC), X-ray fluorescence (XRF) spectrometry and scanning electron microscopy with energy-dispersive X-ray analysis (SEM–EDX), and tin was identified in the scarlet woollen samples (see the appendix for details of the analytical methods).[31]

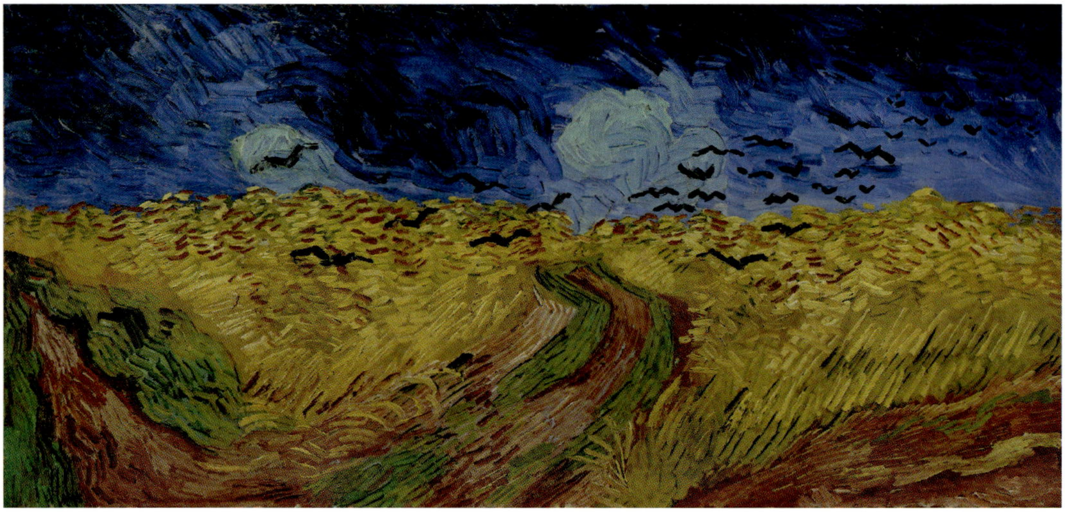

Figure 4 Vincent van Gogh, *Wheatfield with Crows*, 1890, oil on canvas, 50.5 × 103 cm, Van Gogh Museum, Amsterdam, inv. no. s0149V1962, F0779. (Photo: Van Gogh Museum, Amsterdam, Vincent van Gogh Foundation)

Drebbel's Scarlet in textiles

In many dyers' handbooks, recipes are given for preparing and dyeing wool and silk with Drebbel's Scarlet, but its use as a dye in textile objects has rarely been confirmed by analysis. However, woollen yarns dyed with cochineal and tin as a mordant were identified in Turkmen weavings dated between the second half of the 17th century and the mid-19th century, described in the volumes entitled *Turkmen Carpets*, published by Jürg Rageth (1953–2017).[32]

Drebbel's Scarlet was also used in Germany. One of the earliest examples of Drebbel's red is found in the dated tapestry marking the 150th anniversary of the Protestant Reformation, woven by Anna Bump, who was born in 1644 in Hennstedt in Dithmarschen, northern Germany. Besides the name of the creator, the date is also woven into the tapestry: '1667D31X', namely 31 October 1667, at the right-hand edge of the last field (Fig. 3). This tapestry, representing the life of Jesus Christ, was produced in Hennstedt, probably for the St Secundus church. It is now owned by the Museum Europäischer Kulturen, Berlin State Museums.

The tapestry is made of wool, silk and unbleached linen. The natural dyes woad or indigo, weld, gallnut, madder, brazilwood and cochineal have been identified by UHPLC (see the appendix). It contains many red shades between scarlet and pink in which the presence of American cochineal (*Dactylopius coccus* Costa, 1829) has been identified. In the scarlet woollen yarns, tin was detected by XRF and SEM–EDX, confirming its use as a mordant.[33]

Drebbel's Scarlet as a pigment

Methods and recipes to prepare tin lakes of cochineal are known in English, French, German and Italian manuscripts and books from the mid-18th century onwards. In 1758, the English apothecary and chemist Robert Dossie (1717–1777) noticed that there was a similarity between the preparation of the cochineal pigment carmine and dyeing, and even earlier authors had wondered about a similar connection.[34] The method is clearly linked with the use of tin salts in the cochineal dye bath to give a scarlet dye.

The substrate is tin oxide derived from the precipitation of the dye with a solution of tin(IV) chloride or tin nitrate. The tin salts are formed by dissolving tin in *aqua regia* giving tin(IV) chloride, or in *aqua fortis* giving soluble tin nitrates; these cause a precipitate in the dye bath, forming a solid scarlet lake.[35] In later French recipe

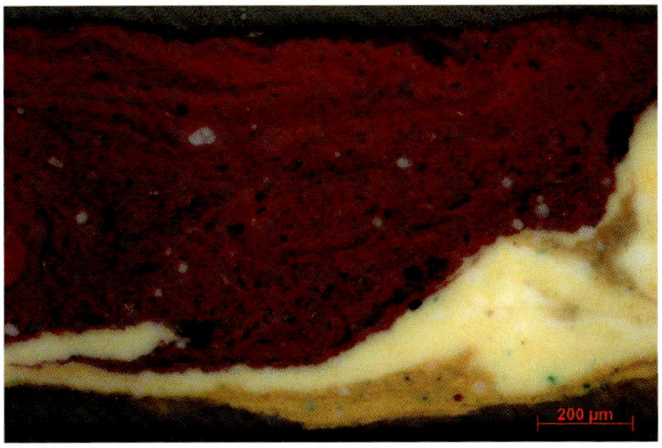

Figure 5 Vincent Van Gogh, *Wheatfield with Crows* (Fig. 4): cross-section of paint containing a non-fluorescing and tin-containing organic red pigment identified as a cochineal lake, photographed in reflected light. (Photo © Muriel Geldof, RCE Amsterdam)

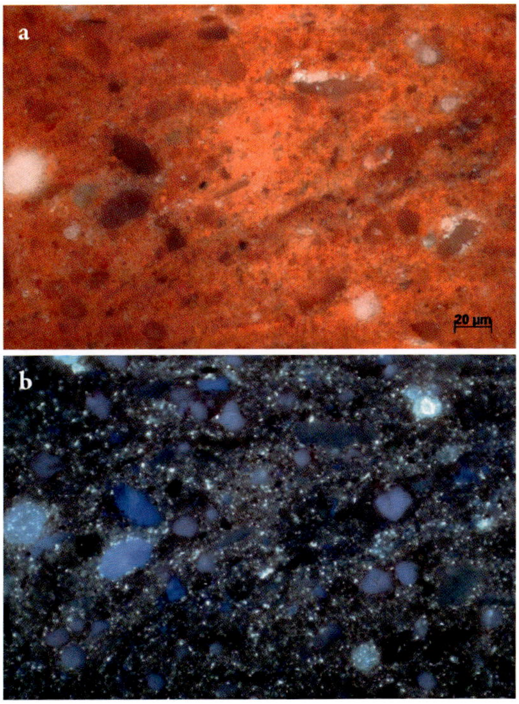

Figure 6 Vincent Van Gogh, *Wheatfield with Crows* (Fig. 4). Detail of the cross-section shown in Figure 5: (a) photographed in reflected light showing whitish translucent starch grains, 10–15 µm in diameter, therefore probably corn starch; (b) in ultraviolet (UV) illumination the starch grains appear bluish and are easily seen. (Photo © Muriel Geldof, RCE Amsterdam)

collections, the name 'Chinese' was associated with the tin(IV) chloride method.[36] The pigments became widely used in the 19th century. Lakes on tin substrates have an intense red colour and are somewhat opaque compared to lakes on a substrate of hydrated alumina, therefore starch was frequently added during the manufacture to achieve a more translucent colour.[37]

Drebbel's Scarlet in painted objects

In the 19th century, artists including Vincent van Gogh (1853–1890) and his contemporaries used tin lakes of cochineal. During the years he lived and worked in France, Van Gogh purchased different types of cochineal paints. In Paris, between 1886 and 1888, his palette evolved into a vivid colour range, including a red lake which he described as 'Carmine is the red of wine, and it's warm, spirited as wine'.[38] Scientific examination shows that he painted with aluminium/calcium-containing and tin-based cochineal lakes. Near the end of his life, Van Gogh moved to the south of France and continued to use various aluminium- and tin-containing carmine lake pigments: a tin lake of cochineal has been identified in 14 paintings dated between 1888 and July 1890.

In the painting *Wheatfield with Crows* (F0779, Van Gogh Museum, Amsterdam) (Fig. 4), painted in July 1890, the tin-containing cochineal lake was mixed with starch grains. The red paint contains granules of starch with a diameter between 10 and 15 µm (Figs 5 and 6), suggesting that they are grains of corn starch. In Paris, Van Gogh had used a slightly different paint containing smaller grains, probably rice starch. Starch was added during the manufacture of the pigment to obtain a more translucent colour.

Van Gogh was not the only 19th-century artist who made use of this type of paint: Emile Bernard and Armand Guillaumin used a similar cochineal paint containing large grains of starch (Figs 7 and 8). Furthermore, Van Gogh's contemporaries Paul Cézanne, Hilaire-Germain-Edgar Degas, Pierre-Auguste Renoir, Claude Monet, Ferdinand-Victor-Eugène Delacroix and Adolphe Monticelli also used tin-based lakes of cochineal.[39]

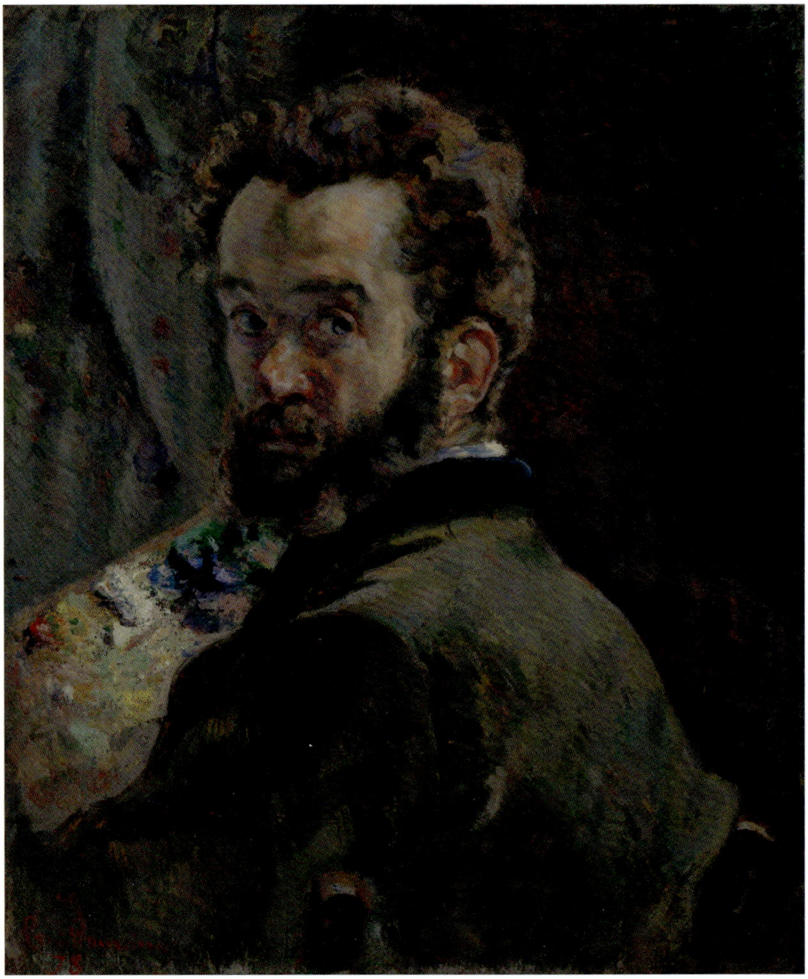

Figure 7 Armand Guillaumin, *Self Portrait with Palette*, 1878, oil on canvas, 60 × 50 cm, Van Gogh Museum, Amsterdam, inv. no. s0228V1962. (Photo: Van Gogh Museum, Amsterdam, Vincent van Gogh Foundation)

Three-part folding screen

A three-part folding screen, each panel measuring 134 × 47 cm, currently resides in a 'Chinese Room' (chinoiserie style), probably since its creation, in the Esterházy Palace in Eisenstadt, Austria. The painting is executed on machine-made paper indicating that the object dates to the second half of the 19th century. The condition of the screen is poor: the front and back papers were originally coloured deep red, but this has faded and is only visible on the back of the panels, in areas previously covered by nails. The red is extremely water-sensitive and very strong tidelines have formed on both the recto and verso in areas damaged by water. On the front of the panels the red has completely faded and now has a transparent brownish layer over the whole surface.

Using UHPLC, the red organic dye carminic acid was identified, together with other characteristic components pointing to the use of the dye insect American cochineal (*Dactylopius coccus* Costa). SEM–EDX analysis detected tin in the red paint layer, making it possible to confirm that the discoloured background of the front papers originally had a red colour given by cochineal.[40]

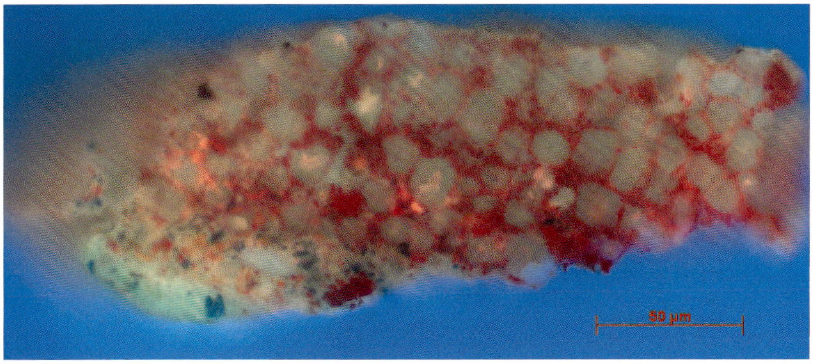

Figure 8 Armand Guillaumin, *Self Portrait with Palette* (Fig. 7): cross-section of paint containing red lake pigment photographed in UV illumination showing large white starch grains. (Photo © Muriel Geldof, RCE Amsterdam)

Scarlet substitutes given by synthetic organic dyes

The term 'scarlet' was formerly confined almost entirely to cochineal on a tin mordant but with the introduction of the synthetic organic dyes, this term was extended to other similar shades. The word 'scarlet' still remained in company and trade names: in the 1931 *Farbstofftabellen* of Gustav Schultz, 'scarlet' appears about 250 times. Drebbel's Scarlet was gradually replaced by cheap manmade synthetic dyes at the end of the 19th century. Important synthetic scarlets include the red acid azo dyes Biebrich Scarlet (C.I. Acid Red 66, 26905) and the Ponceaus given by the 2-naphthol-disulfonic acids R and G. This dye class was discovered in 1878 by the German chemist Heinrich Baum (1849–1923), who worked at the factory of Farbwerke Meister, Lucius & Brüning in Frankfurt Hoechst. Three Ponceau dyes give a scarlet colour, namely Ponceau 2G (C.I. Acid Orange 14, 16100), Ponceau 2R (Scarlet 2R; C.I. Acid Red 26, 16150) and Cochineal Red A (Scarlet 4R; C.I. Acid Red 18, 16255). In Britain, scarlet broadcloth for army officers' uniforms was still dyed with cochineal and tin chloride until 1952, when it was finally replaced by a synthetic dye.[41] By the end of the 19th century there was an outcry that artists were kept in ignorance of the materials they were using and that many colours were fugitive – even in 1930, Martin Fischer warned against the use of many early synthetic dyes, including Scarlet red.[42]

Conclusion

Cornelis Jacobszoon Drebbel was one of the first scientists to put the motto 'Science pays' into practice: no fewer than nine inventions are attributed to him, among which is a scarlet made by mordanting cochineal with a solution produced by dissolving granulated tin in nitric acid. Drebbel's Scarlet is associated with his partners and relations by marriage, the Kuffler brothers. Information concerning his Scarlet are given by Drebbel's contemporaries and secondary texts, but its invention and year of discovery remain in doubt. Although Drebbel's dye-works was established in 1607, the only clear citation on the colour is that from the savant Nicolas-Claude Fabri de Peiresc following his meeting with Johannes Sibertus Kuffler, who showed him specimens of dyed scarlet, in 1627. It cannot be excluded that Drebbel invented or improved the technique before 1627. It is also unclear whether or not Drebbel really did apply a tin salt as a mordant or used a pewter vessel, or both. Literature dated after Drebbel's death does not clarify either point but dyers noticed that pewter vessels also dyed textiles a scarlet colour.

Many dyers' handbooks give recipes for dyeing with Drebbel's Scarlet, but as a dye it is rarely identified in textile objects. However, its presence has been confirmed in Turkmen weavings dated between the second half of the 17th century and the mid-19th century. An earlier example of Drebbel's red is found in the scarlet yarns of the tapestry marking the 150th anniversary of the Protestant

Reformation, woven by Anna Bump in Hennstedt in Dithmarschen, northern Germany, in 1667.

Methods and recipes to prepare cochineal pigments as tin lakes are known from the mid-18th century onwards and the pigments became widely used in the 19th century. The lakes have an intense, rather opaque red colour, therefore starch was frequently added during the manufacture to produce a more translucent colour. Vincent van Gogh and his contemporaries used tin lakes of cochineal mixed with different sorts of starch. The introduction of synthetic organic dyes in the late 19th century caused a gradual decline in the use of Drebbel's Scarlet, which was gradually replaced by the acid azo dyes Biebrich Scarlet and the Ponceaus.

Appendix

UHPLC–PDA

The extraction of the sample consists of two steps. First 50 µL dimethyl sulfoxide (DMSO) is added and the mixture heated for 10 min at 70 °C in a water bath. The first extract solution is pipetted into an insert vial and kept in the dark. Then 35 µL of a solution of two parts concentrated hydrochloric acid (36%), one part water (Millipore) and one part methanol (LC-MS grade) are added to the sample and heated for 10 min at 100 °C in a water bath. The sample is then evaporated under a gentle nitrogen flow. When dry, the sample is dissolved in the earlier obtained DMSO extraction (50 µl). The chromatography apparatus used is a Waters Acquity UPLC H-Class System (UHPLC). The column used is a Waters 15 cm ethylene bridged hybrid (BEH) column with pre-filter. The gradient of three eluents, A: 10% methanol in water, B: 100% methanol and C: 1% formic acid, is managed automatically and 5 µL of the extracted sample is automatically injected. A photodiode array detector (PDA), with wavelength 200–800 nm is used to detect the organic colorants.

XRF

The element tin is measured in the samples by energy-dispersive micro-X-ray fluorescence, model ArtTAXPro apparatus (Firma Bruker) with a molybdenum tube (30 W), nitrogen-free SDD detector and a polycapillary lens (80–100 µm spot). The measuring surface has a diameter of approximately 0.1 mm.

SEM–EDX

An FEI NovaNano450 FEG scanning electron microscope with low vacuum is used to examine the samples. The chemical composition of the material is measured using a Thermo Scientific SDD EDX detector with ThermoFisher Pathfinder software.

Editor's note and authors' addresses

*Author for correspondence: m.de.keijzer@a1.net. At the time the paper was submitted the author was employed at the Cultural Heritage Agency of the Netherlands. Paper received 13 March 2018; revised version 23 March 2019. This paper was presented at *Dyes in History and Archaeology* 36, Hampton Court, UK, 26–28 October, 2017.

[1] Cultural Heritage Agency of the Netherlands, Department Conservation and Restoration, 1071 ZC, Amsterdam, The Netherlands.

Notes and references

1. Snelders, H.A.M. (1993) *De geschiedenis van de scheikunde in Nederland, Van alchemie tot chemie en chemische industrie rond 1900*, part 1, Delft, pp. 42–3.
2. Tierie, G. (1932) 'Cornelis Drebbel (1572–1633)' (PhD thesis, University of Leiden) Leiden, Amsterdam, H.J. Paris, pp. 3–4, 107; Jaeger, F.M. (1922) *Cornelis Drebbel en zijne tijdgenooten*, Groningen, pp. 14–15, 119–23; Harris, L.E. (1961) *The Two Netherlanders*, Leiden, pp. 127–31.
3. Tierie 1932 (see note 2 above), p. 44; Jaeger 1922 (see note 2 above), pp. 135, 137–8; Harris 1961 (see note 2 above), p. 219.
4. Tierie 1932 (see note 2 above), pp. 5–8; Jaeger 1922 (see note 2 above), pp. 39–42, 99–104; Harris 1961 (see note 2 above), pp. 141–8.
5. Tierie 1932 (see note 2 above), pp. 61, 112–13; Jaeger 1922 (see note 2 above), pp. 46–7; Harris 1961 (see note 2 above), pp. 166–7.

6. Bodleian Library, Oxford, MS Rawlinson B. 158, pp. 174–5; Tierie 1932 (see note 2 above), pp. 12–13; Jaeger 1922 (see note 2 above), pp. 29–30, 107.
7. Harris 1961 (see note 2 above), p. 217.
8. Bruce, J. and Hamilton, W.D. (eds) (1871) *Calendar of State Papers Domestic: Charles I, 1638–9; Charles I – volume 404: December 1–31, 1638*, London, p. 176; Shaw, W.A. (ed.) (1911) *Letters of Denization and Acts of Naturalization for Aliens in England, 1603–1700*, Lymington, p. 59; Jaeger 1922 (see note 2 above), pp. 29–30, 49.
9. Jaeger 1922 (see note 2 above), p. 49.
10. Tierie 1932 (see note 2 above), p. 78, citing Leidsch Gemeente Archief, No. 444, Onterman 179, 23 September 1651. The Regionaal Archief Leiden, formerly the Gemeentearchief Leiden, is now part of Erfgoed Leiden en Omstreken – Heritage Leiden and Surroundings; the full archive reference is inv. no. 179, Inventory of the Old Notarial Archives of Leiden, 1564–1811, access no. NL-LdnRAL-0506, XXXVII, Archives of notary public Kaerl Outerman (not Onterman), 1629–1669: 444 Minutes Deeds Notary Kaerl Outerman, 1651. Available at: https://www.erfgoedleiden.nl/collecties/archieven/archievenoverzicht/details/NL-LdnRAL-0506 (accessed 9 January 2021). The relationship of this Jacob Kuffler to the three Kuffler brothers is unknown.
11. Jaeger 1922 (see note 2 above), pp. 47–8, 113–15; Gilden- en Officiantenboek der Stad Arnhem, No. 56, Fol. 281, 282, Oud-Archief te Arnhem (now part of the Gelders Archief), 21 Februarij 1654; Tierie 1932 (see note 2 above); p. 78; Harris 1961 (see note 2 above), p. 218.
12. Jaeger 1922 (see note 2 above), pp. 47–50: Family archive of F.C. van der Meer van Kuffeler, 's-Gravenhage, p. 114 (now the archives of Mej. V.C. van der Meer van Kuffeler, in the National Archives at The Hague, 2.14.20 Inventaris van het archief van het Ministerie van Onderwijs, Kunsten en Wetenschappen: Afdeling Kabinet, 1918–1940); Tierie 1932 (see note 2 above), p. 78; Harris 1961 (see note 2 above), p. 218; Leidsch Gemeente Archief, Getuigenissen, X, 237 verso, 25 Juli 1652. The Gemeentearchief Leiden is now part of Erfgoed Leiden en Omstreken – Heritage Leiden and Surroundings; see Inventory of the City Archives of Leiden, (1253) 1574–1816 (1897), access no. NL-LdnRAL-0501A; it has not been possible to trace the document further.
13. Tierie 1932 (see note 2 above), p. 78; Jaeger 1922 (see note 2 above), pp. 47–8; Harris 1961 (see note 2 above), p. 217.
14. Boyle, R. (1744) *The Works of the Honourable Robert Boyle ... To which is prefixed the life of the author*, ed. T. Birch, 5 vols, London, vol. III, pp. 139, 174; Boyle, R. (1725) *The Philosophical Works of the Honourable Robert Boyle, Esq.; abridged, methodized, and disposed under the general heads of Physics, Statics, Pneumatics, Natural History, Chymistry, and Medicine* ..., ed. P. Shaw, 3 vols, London, vol. I, p. 136; II, pp. 101, 105; Evelyn, J. (1818) *Memoirs illustrative of the life and writings of John Evelyn, comprising his Diary, from ... 1641 to 1705–6, and a selection of his familiar letters* ..., ed. W. Bray, London, p. 370; Hoffmann, F. (1748) *Opera Omnia: Observationum-physico-chymicarum*, Geneva, p. 446; MS Rawlinson B. 158. (see note 6 above), pp. 174–5; Beckmann, J. (1846) *A History of Inventions, Discoveries and Origins*, 4th edn, 2 vols, London, vol. 1, pp. 402–3.
15. Beckmann 1846 (see note 14 above).
16. Tierie 1932 (see note 2 above), pp. 35, 77, citing Peiresc, N.C. Fabri de (n.d) Manuscripts, Bibliothèque Inguimbertine, Carpentras, f. 411; Jaeger 1922 (see note 2 above), p. 130.
17. Tierie 1932 (see note 2 above), p. 80; Petty, W. (1667) 'An Apparatus to the History of the Common Practices of Dying', in T. Sprat, *The History of the Royal-Society of London for the Improving of Natural Knowledge*, London, pp. 284–301; 'General Observations upon Dying', pp. 302–6. Available at: https://quod.lib.umich.edu/e/eebo/A61158.0001.001/1:10.14?rgn=div2;view=fulltext (accessed 2 February 2019). The quotations are from pp. 288, 298 and 304.
18. Boyle 1725 (see note 14 above), p. 105; Tierie 1932 (see note 2 above), p. 81.
19. Kunckel, J. (1679) *Ars vitraria experimentalis, oder Vollkommene Glasmacher-Kunst*, 2 vols, Frankfurt and Leipzig, vol. l, p. 185; Tierie 1932 (see note 2 above), pp. 79–80, 118.
20. Hoffmann 1748 (see note 14 above), p. 446; Tierie 1932 (see note 2 above), p. 79.
21. *Le Teinturier parfait, ou Instruction generale et nouvelle pour la teinture des laines et manufactures de laine* (1708) Leyden, pp. 9, 170; Tierie 1932 (see note 2 above), p. 79.
22. Tierie 1932 (see note 2 above), p. 82, citing Leidsch Gemeente Archief. Notaris Outerman, 3 April 1648. This is now to be found at Erfgoed Leiden en Omstreken, inv. no. 179, Inventory of the Old Notarial Archives of Leiden, 1564–1811, access no. NL-LdnRAL-0506, XXXVII, Archives of notary public Kaerl Outerman, 1629–1669, no. 441, Minutes Deeds Notary Kaerl Outerman, 1648. Available at: https://www.erfgoedleiden.nl/collecties/archieven/archievenoverzicht/details/NL-LdnRAL-0506 (accessed 9 January 2021).
23. I.K.C. (1709) *Der curieuse und vollkommene Kunst-Faerber, aufgeführet in einem gantz-neu heraus gegebenen und in zwey Theilen abgefasten Kunst-Buch* ..., Nuremberg, p. 82; Tierie 1932 (see note 2 above), pp. 80, 118.
24. Tierie 1932 (see note 2 above), pp. 35, 81–2, citing 'Collections of Augustus Kuffeler', University

Library, Cambridge, MS Ll.v.8, part IV (V), p. 5; Jaeger 1922 (see note 2 above), pp. 93–4, 135–7; Harris 1961 (see note 2 above), pp. 218–20.

25. Harris 1961 (see note 2 above), p. 193.
26. MacCulloch, J.R. (1837) *Handbuch für Kaufleute oder Uebersicht der wichtigen Gegenstände des Handels und Manufakturwesens, der Schifffahrt und der Bankgeschäfte, mit steter Beziehung auf National-Oekonomie und Finanzen*, supplement vol., Stuttgart and Tübingen, p. 880.
27. Tierie 1932 (see note 2 above), pp. 32, 79.
28. Phipps, E. (2010) *Cochineal Red: The Art History of a Color*, New York, Metropolitan Museum of Art, pp. 33–4; Tierie, 1932 (see note 2 above), p. 79. On the book itself, see Kirby, J., van Bommel, M. and Verhecken, A. with Spring, M., Vanden Berghe, I., Stege, H. and Richter, M. (2014) *Natural Colorants for Dyeing and Lake Pigments: Practical Recipes and their Historical Sources*, London, pp. 42–4.
29. Tierie 1932 (see note 2 above), pp. 79, 117–18, quoting H. Havard, *Dictionnaire de l'ameublement et de la d*écoration *depuis le XIIIe siècle jusqu'à nos jours*, 4 vols, Paris, vol. II, col. 251 (date of edition used unknown).
30. Bemiss, E. (1973) *The Dyer's Companion*, New York, Dover Publications, Inc., pp. 79, 184–220; Bronson, J. and Bronson, R. (1977) *Early American Weaving and Dyeing: The Domestic Manufacturer's Assistant and Family Directory in the Arts of Weaving and Dyeing*, New York, Dover Publications, Inc., pp. 169–84; Partridge, W. (1973) *A Practical Treatise on Dying of Woollen, Cotton, and Skein Silk with the Manufacture of Broadcloth and Cassimere including the most improved Methods in the West of England*, Edington, Wilts., Pasold Research Fund Ltd., pp. 184–9.
31. Scientific research carried out by the Cultural Heritage Agency of the Netherlands.
32. Rageth, J. (2016) *Turkmen Carpets, A New Perspective*: *An Interdisciplinary Study based on Radiocarbon Dating, Dye, Mordant, and Technical Analyses, as well as Historical and Art Historical Sources*, 2 vols, Basel, vol. 1, pp. 324–5; Vanden Berghe, I. (2016) 'Appendix III: Tables 11–14, organic and inorganic mordant analysis (HPLC- and SEM-element analysis) tin mordant in Turkmen Weavings', in *op. cit.*, pp. 345–8, esp. pp. 345–6.
33. Neuland-Kitzerow, D., Binroth, C. and Joram, S. (2017) *Anna Weaves Reformation: A Tapestry and its Stories*, Husum, Verlag der Kunst Dresden, pp. 9–24, 35–43 (Schriftenreihe Museum Europäischer Kulturen, vol. 20; also in German); Joram, S., Proaño Gaibor, A. Néss, Marten, S., Pausewein, R-R., Schwerdtfleger, S., de Keijzer, M., Hofmann-de Keijzer, R. and Reiche, I. (2017) 'Holländischer Scharlach im Reformationsteppich der Anna Bump aus dem Jahre 1667', *Berliner Beiträge zur Archäometrie, Kunsttechnologie und Konservierungswissenschaft* 25, pp. 169–88.
34. Dossie, R. (1758) *The Handmaid to the Arts*, 2 vols, London, vol. I, pp. 54–5; Kirby, J., Spring, M. and Higgitt. C. (2007) 'The technology of eighteenth- and nineteenth-century red lake pigments', *National Gallery Technical Bulletin* 28, p. 71.
35. Le Pileur d'Apligny, [P.A.] (1776) *L'Art de la teinture des fils et étoffes de coton ...*, Paris, p. 54.
36. Lefort, J. (1855) *Chimie des couleurs pour la peinture à l'eau et à l'huile*, Paris, p. 185.
37. Kirby *et al.* 2007 (see note 34 above), pp. 69–95; Kirby *et al.* 2014 (see note 28 above), pp. 78, 83–4, 101.
38. Letter from Vincent van Gogh to Theo van Gogh, Antwerp, Monday, 28 December 1885, *Vincent van Gogh – The Letters*, ed. L. Jansen, H. Luijten and N. Bakker, Amsterdam 2009 (online version 2020), no. 550. Available at: http://vangoghletters.org/vg/letters/let550/letter.html (accessed 10 January 2021).
39. Geldof, M., Megens, L. and Salvant, J. (2013) 'Van Gogh's palette in Arles, Saint-Rémy and Auvers-sur-Oise', in M. Vellekoop, E. Hendriks, L. Jansen, M. Geldof and A. de Tagle (eds), *Van Gogh's Studio Practice*, New Haven and London, pp. 247–9; Kirby *et al.* 2007 (see note 33 above), pp. 90–93.
40. Steinbichler, M. (2020) 'Scharlachrotes Blass – Untersuchung, Konservierung und Restaurierung eines chinoisen Paravants (2. Hälfte 19. Jhdt.) aus der Sammlung Esterházy' (thesis, Academy of Fine Arts Vienna).
41. Edmonds, J. (1999) *The History and Practice of Eighteenth Century Dyeing*, Little Chalfont, pp. 39–41 (Historic Dyes Series No. 2).
42. Fischer, M. (1930) *The Permanent Palette*, New York (National Publishing Society), pp. 48–9.

Fake Lakes? The Use of Starch and other Adsorbent Substrates for Synthetic Dyes in Paint Manufacture

Eva Eis

Abstract In the late 19th and early 20th century, questions were raised about adulterations in painting materials. Paint manufacturers were accused of selling cheap substrates dyed with aniline dyes instead of lightfast mineral pigments. But if no mordant was used for the precipitation of a dye, can those pigments really be considered as lakes? Production of such colours was common practice in pigment manufacture, as documented in the literature of the time. Recipes for making pigments by dyeing starch, kaolin or green earth with synthetic dyes also appear in handwritten 19th-century sources and reproduction of some of these colours has been carried out. Manufacture of these 'lakes' was simple and they provided cheap substitutes with bright colours. Although their stability led to vigorous debates between manufacturers and users, 'fake lakes' were produced at least until the first half of the 20th century.

Introduction

In 1895, an anonymous author complained about 'faked colours' in *Dingler's Polytechnic Journal*, referring to an analysis carried out by the Research Laboratory for Painting Techniques in Munich,[1] which had shown that several pigments – especially materials used in decorative painting – had been 'adulterated in an irresponsible manner'. Colours that were called 'Wall Green', 'Lime green', 'Fashion blue' or 'Ultramarine replacement' had turned out to be clay or baryte, dyed with aniline dyes, instead of lightfast mineral pigments. The author complained that some of these colours faded within a few hours when exposed to light, whereas genuine ultramarine colours were lightfast and weatherproof. He warned authorities, architects and builders, painters and house owners to be aware of such colours.[2]

In subsequent years a debate on the adulterations in painting materials ensued. Opponents in this discussion were artists and decorative painters on one side and representatives of the paint industry on the other.[3] Manufacturers were accused of adulterating pigments and binders and selling cheap mineral substrates dyed with aniline dyes, enhancing the colours of earth pigments or replacing traditional pigments with completely different materials without labelling these products as imitations. Painters and artists lacked the knowledge to analyse these materials and often damage revealed itself only when the work was finished. In the early 20th century the demand for inexpensive colours had grown; paint manufacturers accused consumers of insisting on low prices and forcing them to produce such low quality pigments. They emphasised, however, that the pigments were still useful for many purposes.[4] Although mainly German sources have been evaluated for this paper, some of them indicate that adulterations in paints and pigments were also reported for other countries.[5] The use of earth pigments and baryte dyed with aniline dyes was not the only matter discussed in these articles, meetings and conferences, however it is mentioned frequently in this dispute. The production of these colours was common practice in pigment manufacture and is often described in the literature of the time.

Real lake or fake?

The literature reveals that not only did opinions on these colours differ but also that the nature of these 'lakes' was totally unclear. In 1880, the renowned German chemist Johann Georg Gentele defined lake pigments as colours made from plant, animal or coal tar dyes and which contained that dye either in pure state or in chemical combination with other substances, or bound by adsorption to white, easily separable substrates.[6] Carl Otto Weber, who published an article on the formation of lake pigments in 1892, did not consider these latter pigments to be lakes. He stated that some organic dyes, mainly basic coal tar dyes, were able to mix with organic and mineral substances to form pigments, which could not be distinguished from genuine lakes at first sight. But simple washing of such lakes would remove most of the dye. In his opinion these 'lakes' were unstable to light and air and had nothing in common with real lakes except their name.[7]

The controversy surrounding these pigments continued into the 20th century: while Josef Bersch considered them makeshift solutions and questioned their stability,[8] Hans Wagner described their advantages and insisted that these lakes were not only lightfast, but also stable to water and lime.[9] Laurenz Bock considered it a phenomenon that by precipitating an unstable dye with green earth, a lightfast pigment could be produced. He admitted that although these pigments were simple to produce, they had not been fully explained scientifically. In his opinion, the fixation on green earth was mainly an adsorption process because it could be done without fully dissolving the dye. Such a method, where the dye was additionally fixed with heat, had even been patented.[10] According to one of these patents, the coloured bodies thus achieved were genuine lake pigments.[11]

In 1929, C.A. Curtis stated that lakes on green and white earths, which were previously assumed to be due purely to adsorption, were now regarded as lake pigments. He thought that the dye formed some kind of salt complex – at least with green earth.[12] A year later, Arthur W.C. Harrison agreed that it was 'generally accepted that the precipitation of basic dyes on green earth is largely due to adsorption', but that 'there is an appreciable percentage of the dyestuff which is assumed to be chemically combined, as it cannot be wholly extracted by ordinary solvent methods'. He also mentioned that 'it is highly controversial as to how far the precipitation is due to adsorption and to chemical combination respectively'. Like Hans Wagner, he claimed that these dyes, if precipitated with resin soap, arsenite or phosphate of alumina, 'give lakes that are considerably more fugitive to light, and in some colours have distinct tendency to bleed'.[13]

In 1928, Wagner wrote that although the nature of these lakes had been thoroughly analysed, it was still unclear. Having done several tests himself, he stated that in precipitates with green earth, a salt complex with the iron(II) silicates in the earth was formed. While in chalk, gypsum, baryte or ochres the dyes were only bound by adsorption, he assumed 'Chemosorption' with white earth, Fuller's earth and kaolin. Finally he concluded that pigments where a salt formation takes place could be considered genuine lakes.[14] However, this referred to green earth lakes and to a lesser degree to white earth – not to substrates with purely adsorptive character and certainly not to pigments made from starch.

Substrates

Several different substrates are described in the literature, but most frequently mentioned are green and white earth, kaolin, baryte, chalk and selenite. For adsorption lakes, Gentele suggested the use of starch in particular.[15] Desirable properties of a substrate were good dispersability and ability to bind to the dye and, importantly, transparency and little or no influence on the colour. Curtis pointed out that substrates were supposed to improve the properties of the lake, such as hiding power and brilliancy, but of course the substrate could just as well make the colour cheaper, less brilliant and less stable.[16]

Green earth, white earth and kaolin

For 'adsorption lakes' green earth was one of the most common substrates used. In fact, several authors stated that its main use was for the production of lake pigments in which the green earth was combined with basic aniline dyes as it binds these without any mordant, the resulting lakes forming paints that are lightfast and stable in limewash.[17] According to Harrison, green earth was the most important precipitating agent for basic dyes:

> Green earth (and white earth) are of outstanding importance, although only used for one big class of colours – lakes from basic dyestuffs. They are indispensable for this valuable and special range of colours. Green earth, and to a rather less degree, white earth, possesses the property of precipitating basic dyestuffs from their solutions.[18]

Lightfastness depended on the type of green earth as well as on the dye itself.[19] Included among the suitable green earths were green earth from Graz in Austria, Veronese green earth, Bohemian green earth and earths from Cyprus or Tirol.[20]

Besides green earth, white earth and kaolin were used for such lakes, although the latter is only mentioned in a few sources. Francis H. Jennison described lakes with china clay as having 'a chalky and bright appearance' and being difficult to wash.[21] An anonymous author recommended it as an alternative for green earth in paper and wallpaper manufacture, and suggested that the inferior fastness of lakes on kaolin may be improved by adding tannins.[22]

The paint manufacturing literature does not really explain what 'white earth' is, but its properties and use were discussed in the journal *Farbe und Lack* between 1924 and 1926. It was considered that this siliceous earth was especially suitable as a substrate for coal tar dyes. When the first earths of this kind were discovered in Germany, even mining experts had difficulties characterising them. Since it was neither clay nor chalk, they called it 'white earth' and sold it under that name.[23] However, when several white earths were analysed and discussed, they were dismissed as being more suitable for cleaning agents or as Fuller's earths than for paint manufacture. The most sought after earth came from deposits near Langquaid and Schierling in Lower Bavaria.[24]

Various methods were used for producing these pigments. The simplest preparation was to wet the earth with water in flat pits or boxes, pour the dye solution over it and mix it thoroughly with shovels. Another technique involved grinding the substrate-dye mixture on an edge-runner mill while sprinkling it with alcohol; this was a rather expensive method, however.[25] According to Curtis, the work had to be done as cheaply as possible and therefore the colours were usually just left to air-dry.[26] The dye could also be added to the dispersed earth in a vat; Harrison considered this a 'more rational wet process, with the dyestuff properly diluted and subsequently well dispersed over the whole of the substratum'.[27] 'Wall greens' or 'Lime greens' were made by dyeing green earth with Malachite Green and various types of that dye.[28] In addition, Victoria Blue, Methylene Blue, Methyl Violet, Fuchsine, Auramine, Chrysoidine, Safranine, Bismarck Brown, Brilliant Brown, Rhodamine and Eosin were mentioned for adsorption lakes.[29] According to Wagner, a wide range of colours could be achieved: green, violet, yellow-brown, red-brown, blue and transparent browns. He even mentions dull reds, which could be used as substitutes for dragon's blood and brazilwood lakes.[30]

Starch

In comparison with the amount of literature on green and white earth as substrates for lake pigments, information on the use of starch is almost non-existent – if referenced at all, it is usually considered an adulterant. It seems as though authors were aware of the use of starch as a substrate but did not mention it. For example, Georg Zerr and Robert Rübencamp listed it among the most used substrates but without any further description;[31] only Gentele pointed out its importance in lake making in 1880. He stated that aniline dyes do not adhere well to mineral substrates and that starch, cotton, cellulose and wool fibres would be the perfect substrates, if only it were possible to disperse them into fine powders.[32] This would certainly be difficult with textile fibres, which left starch as the finest, most adsorbent substrate available.

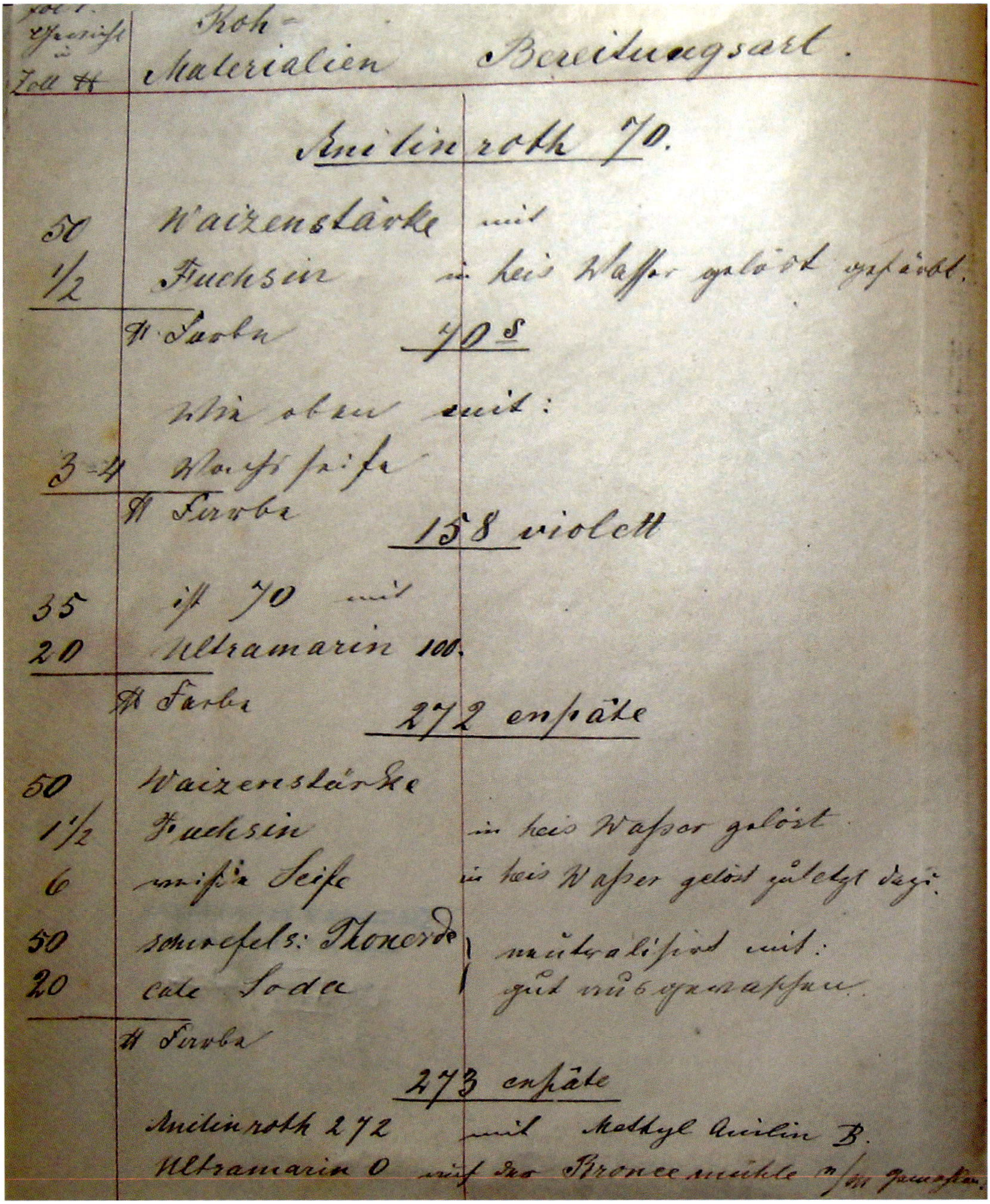
Roh-Materialien | Bereitungsart.

Anilinroth 70.

50 Waizenstärke mit
1/2 Fuchsin in heiß Wasser gelöst gefärbt.

Farbe 70s

wie oben mit:
3-4 Wachsseife

Farbe 158 violett

35 ist 70 mit
20 Ultramarin 100.

Farbe 272 en pâte

50 Waizenstärke
1 1/2 Fuchsin in heiß Wasser gelöst.
6 weiße Seife in heiß Wasser gelöst zuletzt dazu.
50 schwefels: Thonerde neutralisirt mit:
20 calc Soda gut aus gewaschen.

Farbe 273 en pâte

Anilinroth 272 mit Methyl Anilin B.
Ultramarin 0 auf der Broncemühle

Figure 1 Recipes for 'Aniline Red' with starch and Fuchsine in the manufacturing book of Wilhelm Sattler. While the first recipe (Aniline Red 70) contains just these two components, the number below, '70s', additionally contains wax soap and '158 violett' is made with Ultramarine blue. The last two recipes (numbers 272 and 273) were made with an alumina substrate and sold as colour pastes – '*en pâte*'. Source: 'Fabrikationsbuch Farbfabrik Sattler', Gemeindearchiv Schonungen G XVII/77, f. 1.

But the use of starch was neither new nor an invention of the 19th century: starch particles have been found in earlier paint samples, mainly in transparent red glazes made with cochineal and brazilwood lakes of the 17th and 18th centuries, although it is not mentioned in the literature of the time.[33] The situation was more or less the same for 19th-century publications: Harrison still suggested its use as an additive or substrate for cochineal and logwood lakes in 1930, but not in adsorption lakes.[34] Nevertheless it is a frequently mentioned additive in historic recipes. The use of starch was certainly well known and was probably simply adapted to the newly invented synthetic dyes. However, the lack of literature on this matter suggests that this was not an adequate method of making lake pigments.

Historical recipes

Instructions for making 'adsorption lakes' can be found in two handwritten 19th-century sources. The first is a recipe collection compiled by Hermann and Otto Wiesel, two brothers who owned a paint factory in Gehren in Thuringia. The collection contains hundreds of recipes for all kinds of pigments and paints, which were collected from different sources between 1888 and 1894. Among these recipes there are approximately 40 instructions for making 'adsorption lakes'.[35] The other example is the manufacturing book of Wilhelm Sattler, the owner of a renowned pigment factory in Schweinfurt. This source only contains a few recipes of interest in this context. On the first page of the book, several instructions for making 'Aniline Red' with starch and Fuchsine (Basic Violet 14, C.I. 42510) can be found (Fig. 1). In addition, there is a recipe for 'Wallgreen', made by precipitating Brilliant Green (probably Acid Green 1, C.I. 42040) with green earth.[36]

A few of these colours were clearly intended for wall paints or decorative painting, such as lime-stable greens, reds or pinks. In addition, imitations of 'Bremen Blue' or 'Schweinfurt Green' can be found.[37] The manufacture of these 'lakes' was relatively simple – the substrate just had to be dispersed in water before the dye solution was added. Most of these recipes simply list the ingredients without an accompanying description as manufacturers usually knew how to proceed. The existing instructions are more or less similar: mineral substrates, such as green earth, red bole, baryte and others, were finely ground before dispersing them in water and adding the dye solution. For pigments that contained starch as an adsorbent substrate, the starch was simply stirred into cold water. The aniline dyes were usually dissolved in hot water, but for starch lakes the solution had to be left to cool down before adding it to the dispersed starch. The mixture was then allowed to rest for a day before filtering and drying. According to Curtis, the yield of these dyes was enormous: it allowed production of pigments which contained only one percent of dye and still had an intense colour.[38] The recipes in the historic collections confirm that the amount of dye needed was relatively small: for example, a 'Bremen Blue imitation' could contain only 1 kg of dye and 140 kg of kaolin.[39] Starch was mainly used as a substrate for Fuchsine, but also for Methyl Violet and a few other dyes. In addition, Fuchsine was used with mineral substrates such as red bole. Green and blue pigments were made with kaolin, baryte, chalk or selenite. Green earth is only mentioned in combination with Brilliant Green or Malachite Green. White earth is not cited in these sources at all, probably because the trading of this material only started in the early 20th century.[40] Some of the dyes were also combined to give certain shades; for example, Auramine was mixed with Brilliant Green for a Schweinfurt green imitation, and Night Green[41] and Methyl Violet were used for making Bremen Blue.[42] Although yellow adsorption lakes are mentioned in the literature, none was found in the historic sources discussed here.

Reproduction of recipes

For the experimental part of this study, five recipes from the Wiesel collection were chosen for reproduction and testing of the stability of the resulting pigments:[43] three of these recipes utilised mineral substrates and two, starch. 'Wall Green Ia' and 'Bremen Blue imit.' were both made with the same dye, Brillantgrün BASF (Basic Green 1, C.I. 42040), which was also sold as 'Solidgrün JJO'. The dye was provided by the Historische Farbstoffsammlung of the Technical University Dresden. 'Wall Green Ia' was produced using baryte, selenite, Bohemian green earth and the green dye (Figs 2 and 3).[44]

The mineral substrates were first dispersed in water and ground in an electric mortar. The dye was dissolved in warm water and added to the substrate. The mixture was ground until homogenous, then filtered and dried. The result was an intense

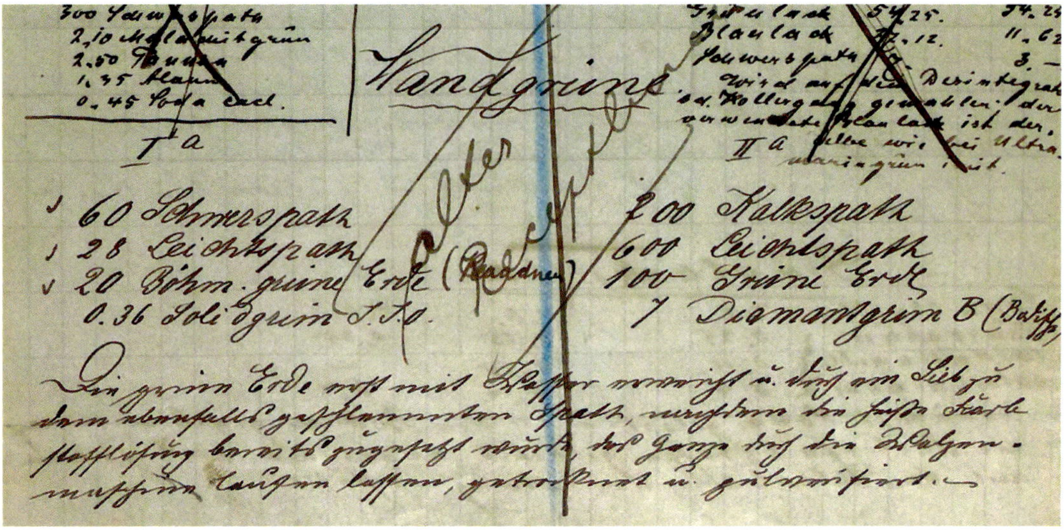
Wandgrüne

I a

60 Schwerspath
28 Leichtspath
20 Böhm. grüne Erde
0.36 Solidgrün J.J.O.

II a

200 Kalkspath
600 Leichtspath
100 Grüne Erde
7 Diamantgrün B (Badisch

Figure 2 The recipe for 'Wall Green' contains two variants, 'Ia' and 'IIa'. Both recipes list mineral fillers and Bohemian green earth, but they use two different types of Malachite Green. While 'Wall Green Ia' was made with 'Solidgrün JJO' (Basic Green 1, C.I. 42040), number 'IIa' was made with 'Diamantgrün B' (Basic Green 4, C.I. 42000). Source: Wiesel Collection, WUL111/004/003, Collection of Kremer Pigmente, Aichstetten.

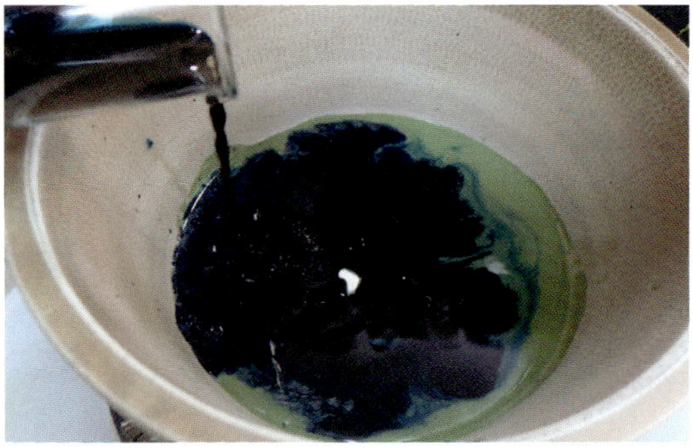

Figure 3 Reproduction of 'Wall Green Ia'. The 'Solid Green' dye is added to the green earth, baryte and selenite mixture before grinding.

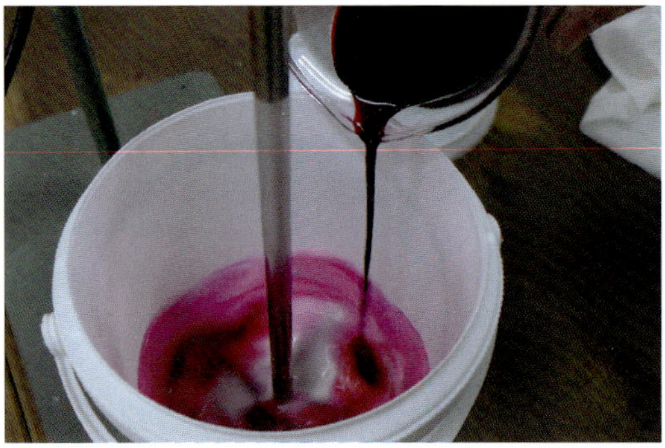

Figure 4 Making 'Brilliant Red' by stirring a Fuchsine solution into the dispersed starch.

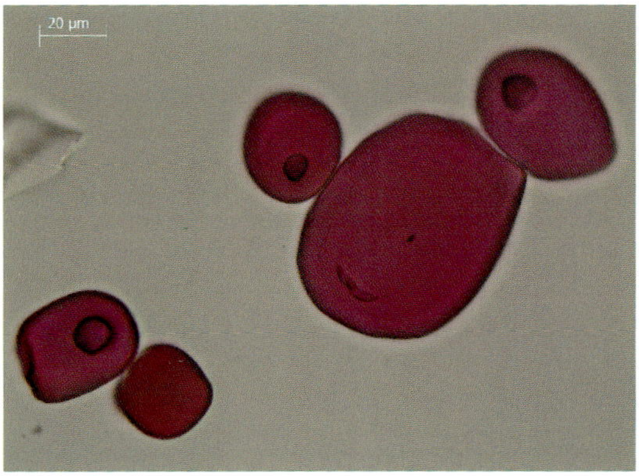

Figure 5 Microscope sample of 'Brilliant Red', showing the dyed potato starch particles.

and very bluish green. 'Burgundy Red' was made the same way by grinding red bole and Caput mortuum and then adding a hot Fuchsine solution. The kaolin-based 'Bremen Blue imitation' was produced by dispersing fine China clay in water. The mixture was stirred continuously while adding a hot solution of Brilliant Green, left to settle overnight, then filtered and dried. The lakes produced from starch, 'Brillant Red' and 'Magenta Carmine', were made in a similar way. For 'Brillant Red' potato starch was stirred into cold water. The Fuchsine was dissolved

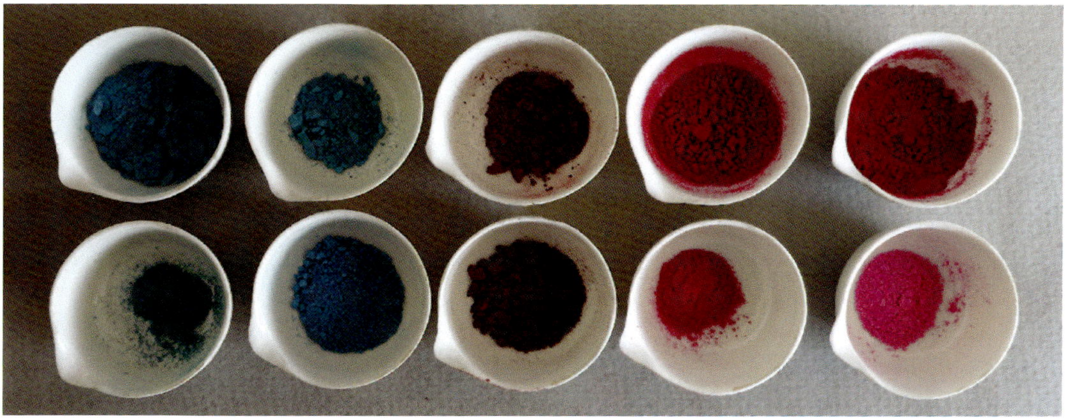

Figure 6 From left to right: 'Bremen Blue imitation', 'Wall Green', 'Burgundy Red', 'Brilliant Red' and 'Magenta Carmine'. Only 'Magenta Carmine' shows a significant colour change before (top) and after washing (beneath).

in hot water; the solution was left to cool down, and then added to the dispersed starch (Figs 4 and 5). The recipe for 'Magenta Carmine' only differs slightly: it is made from wheat starch instead of potato starch and a small amount of acetic acid has to be added. Again, the colours were left to stand overnight before filtering and drying.

As mentioned earlier, several authors questioned the stability of these lakes, especially their lightfastness and tendency to bleed. According to Jennison and Weber, the colour of these lakes could 'be entirely removed by washing'.[45] Several tests were carried out to determine whether or not this was really true.

To test for bleeding, the pigments were washed several times. This was accomplished by dispersing a small amount of each pigment in water and leaving it to settle. The next day, the wash water was poured off and fresh water added. This process was repeated five times with each pigment. The water from each wash was filtered to see whether the remaining colour was insoluble pigment residue or dye that had redissolved. Bleeding of 'Bremen Blue imitation', 'Wall Green Ia' and 'Burgundy Red' could not be detected with this method. After drying, the colours seemed to be more or less unchanged. The wash water of the starch lakes was pink, but appeared to become less intense after several washings. But it was not just excessive dye which had been removed by the washing process: after drying, although 'Brilliant Red' did not look very different, 'Magenta Carmine' had become much paler (Fig. 6). Since some authors indicated that these pigments were easily detected due to their solubility in alcohol,[46] a small amount of pigment was dispersed in ethyl alcohol – only the dye in 'Magenta Carmine' dissolved immediately,

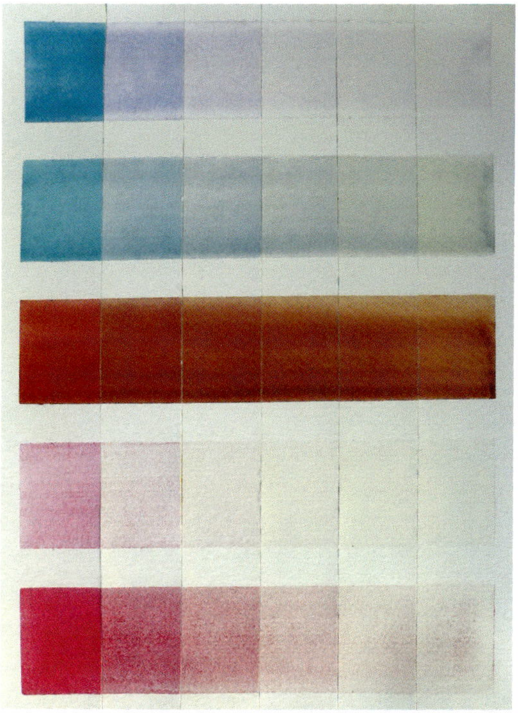

Figure 7 From top to bottom: 'Bremen Blue imitation', 'Wall Green', 'Burgundy Red', 'Brilliant Red' and 'Magenta Carmine', exposed to light for five weeks.

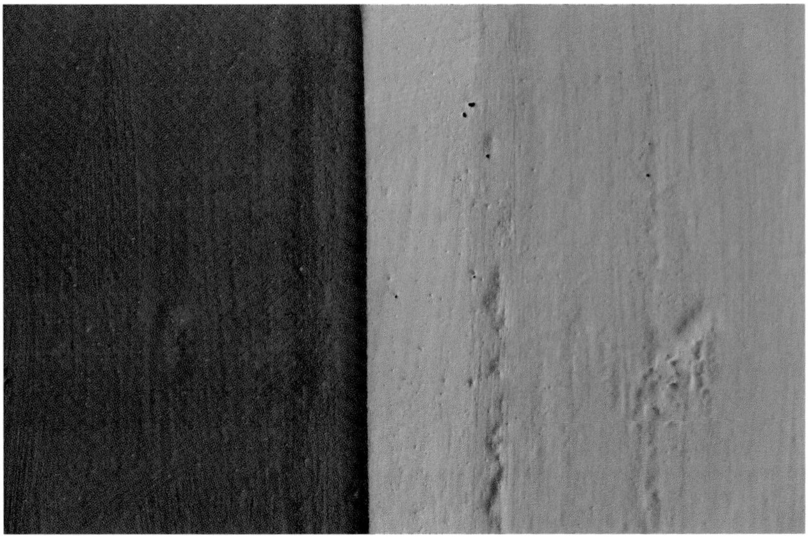

Figure 8 Lime paints with 'Burgundy Red' (left) and 'Wall Green' (right) after drying.

forming a bright pink solution. By the next day, the dyes used for 'Burgundy Red', 'Bremen Blue imitation' and 'Wall Green Ia' had also dissolved in the alcohol. Surprisingly, 'Brilliant Red' remained the most stable.

To test for lightfastness, watercolour paints were made, applied on paper and exposed to light. All samples showed significant colour changes within the first few days (Fig. 7). But although their lightfastness was rather poor, there were differences between the colours. Of the mineral substrate-based pigments, 'Bremen Blue imitation' and 'Burgundy Red' faded quickly, while 'Wall Green Ia', the lake made with green earth, seemed to be more stable. 'Brilliant Red' faded much faster than 'Magenta Carmine'. The pigments were also tested for their compatibilities in egg tempera and oil. All mineral substrate-based pigments were easy to grind and showed no problems with any of the binding media; even the oil paints dried quite well. In oil, 'Wall Green Ia' and 'Bremen Blue imitation' were as transparent and similar in hue to Viridian Green (hydrated chromium oxide, Pigment Green 18, C.I. 77289). The two pigments that contained starch as a substrate were more difficult to work with: 'Brilliant Red' was smeary and hard to grind, and graininess remained visible in the painted samples. 'Magenta Carmine' performed only slightly better, but was easier to apply and had a more vivid colour.

Because a remark in the original recipes indicated the use of 'Wall Green Ia' and 'Burgundy Red' in limewash, these two colours were tested for their stability to lime. Both pigments were mixed with slaked lime and painted onto a brick plate that had previously been grounded with limewash. Although the wet paint of 'Burgundy Red' was a dark wine-red colour that was easy to apply, the dried paint was a disappointing pale rose colour, close to an ordinary iron oxide pigment, and the Fuchsine had bled into the white parts of the limewash, colouring them pink. The 'Wall Green Ia' pigment remained grainy: it did not mix well with the slaked lime and the colour was reminiscent of a pale green earth, but during drying the dull colour became a light, vibrant turquoise green, which showed absolutely no bleeding (Fig. 8). The colours, especially 'Wall Green Ia', seem to be more lightfast than in watercolour, but both colours turned out to be utterly unstable outdoors.

Conclusion

Curtis stated that 'it is not an art to make these lakes'[47] and this is certainly true. Reproducing these colours using the historic recipes is simple and straightforward. The tests confirmed that lightfastness is indeed very poor, although the stability to lime and other binding media seems actually quite good. Apart from 'Magenta Carmine', most of the 'fake lakes' even proved surprisingly stable to washing. Nevertheless, it was Jennison's opinion that these pigments were 'not of a good class'[48] and there can be no doubt that they provided cheap substitutes with bright colours. They were sold in quite large amounts at low prices and were mainly used for purposes where colour intensity was more important than fastness properties – for example, in wall paints or decorative painting as well as for the production of toys and wallpapers.[49] They may also have been used for other applications: Wagner, for example, considered them suitable for use as studio pigments or pastel chalks for artists, for lithography and many other purposes.[50] And even though their stability had already been questioned by the end of the 19th century, 'fake lakes' continued to be produced at least into the first half of the 20th century.

Editor's note and author's address

Address for correspondence: Kremer Pigmente GmbH & Co. KG, Aichstetten, Germany; eis@kremer-pigmente.de. Paper received by 13 March 2018; revised version received by 16 September 2018. This paper was presented at *Dyes in History and Archaeology* 35, Pisa, Italy, 6–8 October 2016.

Acknowledgements

I would like to thank Dr Georg Kremer from Kremer Pigmente for his support, and the Historische Farbstoffsammlung of the Technische Universität Dresden (TU Dresden) for providing the historic sample of 'Brillantgrün BASF = Solidgrün JJO C. I. 42040' used in the experiments.

Notes and references

1. Versuchsanstalt der deutschen Gesellschaft zur Beförderung rationeller Malverfahren.
2. Anonymous (1895) 'Farbenfälschungen', *Dingler's Polytechnisches Journal*, 296, p. 24. Available at: http://dingler.culture.hu-berlin.de/article/pj296/mi296mi01_3 (accessed 19 February 2017, checked 25 January 2021).
3. This dispute has been thoroughly researched. For further information see: Kinseher, K. (2014) *"Womit sollen wir malen?" Farben-Streit und maltechnische Forschung in München. Ein Beitrag zum Wirken von Adolf Wilhelm Keim*, Munich, pp. 136–49.
4. Keim, A.W. (1904) 'Zum Kapitel Farbenfälschung und Farbenschlendrian', *Technische Mitteilungen für Malerei* Jg. 20, No. 17, 1 March 1904, p. 190. All issues of this journal are available at: www.technischemitteilungen.com (accessed 19 February 2017, checked 25 January 2021).
5. Uebele, C.B. (1904) 'Über die Bekämpfung der Farbenfälschungen in Amerika', *Technische Mitteilungen für Malerei* Jg. 20, No. 19, 1 April 1904, pp. 218–19; 'Gegen die Farbenfälschung in Frankreich', *Technische Mitteilungen für Malerei* Jg. 9, Nos. 138 & 139, 1 February 1892, pp. 19–20; Kinseher 2014 (see note 3 above), p. 74.
6. Gentele, J.G. (1880) *Lehrbuch der Farbenfabrikation*, 2nd edn, Braunschweig, pp. 400–410.
7. Weber, C.O. (1892) 'Untersuchungen über die Bildung der Farblacke', *Dingler's Polytechnisches Journal* 283, pp. 158–62. Available at: http://dingler.culture.hu-berlin.de/article/pj283/ar283039 (accessed 19 February 2017, checked 25 January 2021).
8. Bersch, J. (1918) *Die Fabrikation der Erdfarben*, 3rd edn, Vienna and Leipzig, pp. 98–9.
9. Wagner, H. (1923) 'Die Verwendungsmöglichkeiten der Kalkfarben', *Technische Mitteilungen für Malerei*, Jg. 39, No. 3, March 1923, pp. 29–30.
10. Bock, L. (1927) *Herstellung von Buntfarben*, Halle (Saale), pp. 55, 67–8.
11. DE000000429484A, 27 May 1926. Available at: https://depatisnet.dpma.de (accessed 9 January 2017, checked 25 January 2021).
12. Curtis, C.A. (1929) *Künstliche organische Pigmentfarben und ihre Anwendungsgebiete*, Berlin, pp. 6–7 and 19.
13. Harrison, A.W.C. (1930) *The Manufacture of Lakes and Precipitated Pigments*, London, pp. 192–3.
14. Wagner, H. (1928) *Die Körperfarben*, Stuttgart, pp. 303, 309.
15. Gentele 1880 (see note 6 above), pp. 400–401.
16. Curtis 1929 (see note 12 above), p. 1.
17. Plessow, G. (1928) *Die Anstrichstoffe. Ein Auskunftsbuch* über *ihre Rohstoffe, Herstellung, Eigenheiten, Verwendung, Bezugsquellen usw.*, Berlin and Leipzig, p. 77; Curtis 1929 (see note 12 above), p. 13.
18. Harrison 1930 (see note 13 above), pp. 182–3, 192–3.
19. Wagner 1923 (see note 9 above), p. 29.
20. Curtis 1929 (see note 12 above), p. 20; Harrison 1930 (see note 13 above), p. 193; Remington, J.S. (1949) *Pigments: Their Manufacture and Properties*, London, p. 168.
21. Jennison, F.H. (1900) *The Manufacture of Lake Pigments from Artificial Colours*, London, pp. 61–2.

22. Anonymous (1926) 'Grünerdelacke für Tapeten und Buntpapier', *Farbe und Lack* No. 11, 10 March 1926, p. 129.
23. MK (1926) 'Kieselerde und Weisserde für Farbstoffe', *Farbe und Lack* No. 5, 27 January 1926, p. 51.
24. Merkel, A. (1924) 'Über kieselsaure Weißerde', *Farbe und Lack* No. 45, 29 October 1924, p. 433.
25. Wagner (1928) (see note 14 above), pp. 278, 322; Curtis (1929) (see note 12 above), pp. 20–21.
26. Curtis 1929 (see note 12 above), p. 20.
27. Harrison 1930 (see note 13 above), pp. 194–5.
28. The term 'Malachite Green' can refer to Basic Green 1 or Basic Green 4 (C.I. 42000 or C.I. 42040). Just like many other dyes, Malachite Green was traded under many different names. Although most of the sources cited in this paper use common trade names, they do not mention a specific type of dye, therefore it is not always possible to confirm to which dye the source refers. In this paper, Colour Index information is only given if the trade name can be identified.
29. Wagner 1923 (see note 9 above), p. 32; Curtis 1929 (see note 12 above), pp. 6–7; Harrison 1930 (see note 13 above), pp. 193, 195; Bock (1927) (see note 10 above), pp. 68–9.
30. Wagner 1923 (see note 9 above), p. 31.
31. Zerr, G. and Rübencamp, R. (1922) *Handbuch der Farbenfabrikation*, 3rd edn, Berlin, p. 699.
32. Gentele 1880 (see note 6 above), pp. 400–401.
33. Richter, M. (2013) 'Art-technological source research and scientific examination: a comparative study on the technique of coloured glazes applied on metal leaf', in E. Emmerling, M. Kühlenthal and M. Richter (eds), *Lüsterfassungen des Barock und Rokoko / Coloured Glazes on Metal Leaf from the Baroque and Rococo*, Munich, pp. 68–9.
34. Harrison 1930 (see note 13 above), pp. 239 and 242.
35. The documents are owned by Kremer Pigmente. The Wiesel collection has been the topic of the author's doctoral thesis at the Technical University of Munich. The complete transcript of the recipe collection has been published in: Eis, E. (2020) *Die "Farben-Recepte" der Firma Heinrich Wiesel. Transkription und Auswertung der Rezeptsammlung eines Farbenfabrikanten aus dem ausgehenden 19. Jahrhundert.* Bad Langensalza. Within the collection, the recipes mentioned here can also be found using the references given in this article.
36. 'Fabrikationsbuch Farbfabrik Sattler', Gemeindearchiv Schonungen G XVII/77, f. 1 and unfoliated page.
37. Both Bremen Blue and Schweinfurt Green are inorganic copper-based pigments. Schweinfurt Green (or Emerald Green) is a copper acetoarsenite; Bremen Blue was probably a copper hydroxide-based pigment. For further information on both pigments see Eastaugh, N., Walsh, V., Chaplin, T. and Siddall, R. (2005) *The Pigment Compendium: A Dictionary of Historical Pigments*, Oxford, pp. 61, 122, 149.
38. Curtis 1929 (see note 12 above), p. 7.
39. Wiesel collection, Notebook 1, p. 32.
40. Plessow 1928 (see note 17 above), p. 221.
41. It is not clear to which dye this recipe refers. It mentions 'Nachtgrün BB276', which is not a known trade name. Weiler-Ter Meer produced a 'Nachtgrün BB' (or 2B), but that dye was discovered in 1899, five years later than this recipe. It is more likely that it refers to a dye which was discovered by Keisser in 1866 and was no longer available in 1891 (the recipe dates from 1894). This dye was also called 'Jodgrün' or 'Vert lumière'. Schultz, G. and Julius, P. (1891) *Tabellarische* Übersicht *der künstlichen organischen Farbstoffe*, Berlin, No. 284; Lehmann, L. (1931) *Farbstofftabellen von Gustav Schultz. Erster Band. Künstliche organische Farbstoffe bekannter Konstitution oder Herstellungsweise. Natürliche Farbstoffe und Farblacke aus Naturfarbstoffen. Anorganische Farbstoffe (Erd- und Mineralfarben)*, Berlin, No. 767.
42. Wiesel collection, HEU304/042/026 or HEU305/088/024 and 033.
43. Wiesel collection, WUL111/004/001; WUL109/086/017; ARZ303/059/003; Notebook 1, p. 32 and Notebook 2, p. 16.
44. This colour had been made for an earlier presentation at the Futurahma Conference in Pisa in June 2016 and was the stimulus for the author to carry out further research on these pigments.
45. Weber 1892 (see note 7 above), pp. 158–62; Jennison 1900 (see note 21 above), p. 61.
46. Bock 1927 (see note 10 above), p. 107.
47. Curtis 1929 (see note 12 above), p. 20.
48. Jennison 1900 (see note 21 above), p. 64.
49. Curtis 1929 (see note 12 above), pp. 20–21.
50. Wagner 1923 (see note 9 above), p. 29.

Treading on Lichen: The Use of Cudbear by a Victorian Carpet Manufacturer

Terry T. Schaeffer[1*] and Laura Maccarelli[1]

Abstract An early Victorian carpet manufacturer, W.H. Worth, used cudbear extensively as suggested by recipes in a sample book he assembled beginning in 1842. About 30% of the recipes list this British lichen dyestuff, but never mention archil, the purple dyestuff prepared from European lichens. Worth used cudbear alone in recipes for purples, and in combination with other dyes for five additional colours – blued crimsons, browns, salmon, greys and drabs. The presence of cudbear in the yarns accompanying some recipes was investigated by analysing selected samples using fibre optic reflectance spectroscopy, excitation-emission and three-dimensional fluorescence spectroscopy, and high-performance liquid chromatography with diode array detection at 560 nm. The results of these studies confirmed that cudbear was the only dyestuff in the purple samples. Even when cudbear was a very minor component, as implied by dye weight in a greyish drab recipe, the lichen dyestuff was clearly detected by all three methods.

Introduction

The purple dyestuff cudbear, patented and heavily promoted by the Scot Cuthbert Gordon in the latter part of the 18th century, was prepared from lichens growing in Scotland.[1] By the mid-19th century, it was manufactured in England as well as Scotland.[2] Over time, cudbear has been distinguished in various ways from archil, also known as orchil, the purple dyestuff prepared from lichens in continental Europe. Cudbear has been described as being obtained from different species of lichens, as being a powder rather than a paste like archil, and as processed using different times, temperature and/or alkalinity during the extraction process.[3] It has also been defined as the English name for archil dye.[4]

Cudbear was listed often in recipes in Victorian dye manuals[5] and was commonly used as a carpet dye.[6] Thus it was not surprising to find cudbear mentioned frequently in an 1842 sample book of recipes and woollen yarns assembled by William Henry Worth, a partner in the English carpet manufacturing firm of Butcher, Worth and Holmes.[7] Archil and orchil are not mentioned at all in the book, which is an excellent resource on dyeing at that time as it contains 185 samples and over 130 recipes.[8] Most of the yarn samples appear to be in fine condition and are still vividly coloured. Cudbear use as implied by Worth's recipes is summarised below.

The question always arises in respect of historic dyers' recipe books with samples as to whether those samples actually represent the recipes that accompany them. Worth's book provided an unusual opportunity to confirm details of cudbear use specifically in wool carpets in early Victorian England. To verify that Worth did use cudbear as indicated by the recipes in his book, three analytical methods were applied to selected yarn samples: non-invasive fibre optic reflectance

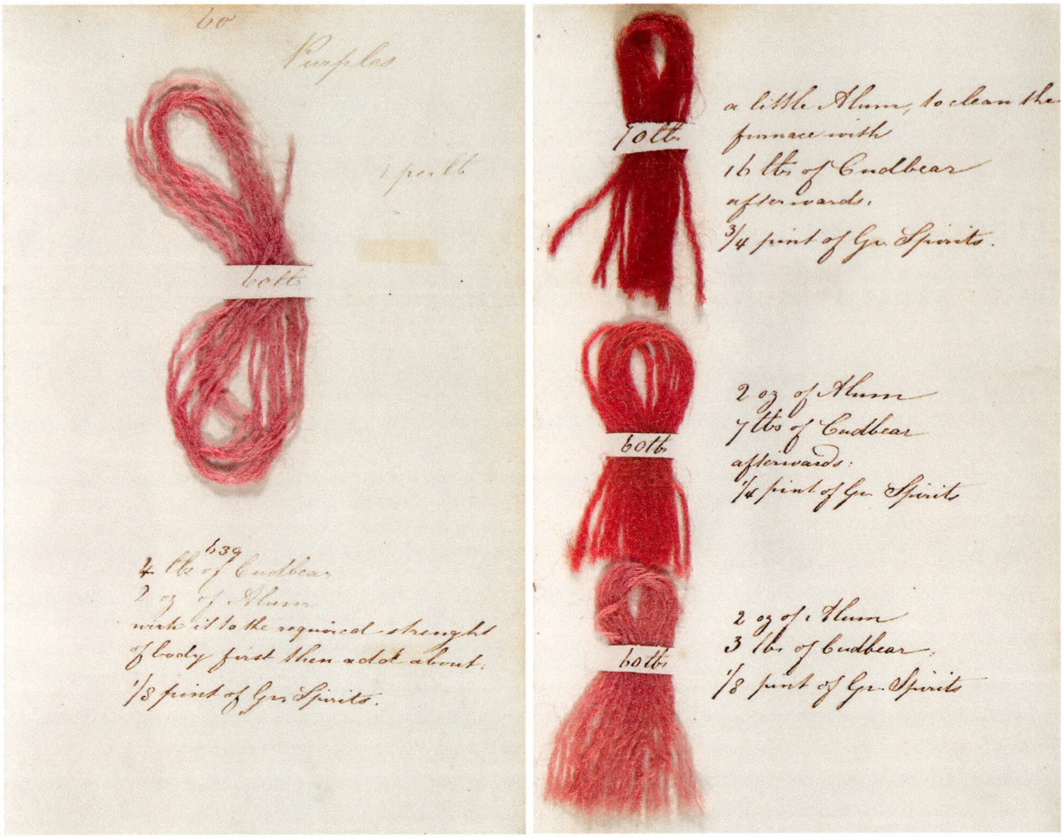

Figure 1 Pages 60 and 61 in the Purple section of the Worth book. Relative amounts of cudbear and grain spirits are varied in the recipes. Gift of Lloyd and Margit Cotsen. (Photo © Museum Associates / LACMA Conservation Center by Yosi Pozeilov)

spectroscopy (FORS); non-destructive spectrofluorimetry; and the micro-destructive procedure of high-performance liquid chromatography with diode array detection (HPLC–DAD). The suitability of each of these methods for detecting cudbear, without having available the means of identifying individual chemical components of the dyestuff, was tested on modern reference materials. These materials included archil-dyed wool yarn and cudbear powder prepared from British lichens in the mid-20th century.

Cudbear use suggested by recipes in the Worth sample book

Worth named the different colour groups represented in his sample book, 'A Collection of Receipts for dying Various Colours made by William Henry Worth begun in October 1842'. He used cudbear in recipes for purple and for five additional colours where it is combined with other dyes. Cudbear is an ingredient in 1 of 2 recipes for salmon, in 2 of 9 recipes for browns, in 3 of 4 recipes for bluing crimson, in 3 of 5 recipes for grey, and in approximately half of about 40 drab recipes. It is also an ingredient in several recipes not assigned to a particular colour group.

Over 40 recipes in the sample book – almost 30% of the total – call for cudbear. This extent of use is similar to the frequency of cudbear-containing recipes included in 19th-century English dye manuals, such as Love and Malony and many others evaluated by Celoria.[9] The amounts of the dye required in Worth's recipes range from 1 ounce (oz) to 16 pounds (lb) (equivalent to 28.35 g and *c*.7.26 kg respectively). The largest amount of cudbear is for dyeing 70 lb of wool purple. Three additional recipes for purples use 3–7 lb of cudbear for 60 lb of wool. No other dye is required by these recipes. Other ingredients in the four purple recipes include alum and grain spirits.[10] The relative amounts of cudbear and the other ingredients vary among the four recipes, resulting in shade variations: generally, the more grain spirits used, the redder the shade (Fig. 1).

In recipes for colours other than purple, cudbear is never the major dye component by weight, although in a few drab recipes it is slightly more than one-third of the total dye weight. Most often cudbear is present as a minor component, contributing a few percent of the total amount of dyes used. In some cases it is listed for darkening a brown or drab colour.

When cudbear is required to blue a crimson-dyed yarn, the only other dye in the recipes is cochineal. In recipes for other colours with cudbear among the dyes listed, fustic is always present. Red dyes often appear in these recipes, usually madder and occasionally redwood. Fustic and madder are the only dyes mentioned more frequently in the book than cudbear. A small number of the recipes that include cudbear do not list any dyeing aids, but many require alum and/or tartar. The blued crimson and the salmon recipes have acidic tin solutions. The brown recipes that include cudbear list copperas (ferrous – iron(II) – sulfate). Some drab recipes requiring cudbear call for the dyeing aid argol, which is most likely a less pure variety of a tartar precipitate from wine casks.[11]

Cudbear use as evidenced by analyses of Worth's yarn samples

Materials

Schweppe yarns

Yarns dyed in the latter half of the 20th century by Helmut Schweppe with archil extract from *Roccella tinctoria* DC. (= *Lecanora tinctoria* (DC.) Czerwiak, recently reclassified as *Roccella phycopsis* Ach.), using 50% of the weight of the wool, were used as reference materials. Schweppe treated sample #300 with ammonia and for sample #647 mordanted with 25% alum and 6% tartar, and an unspecified amount of acetic acid was added.

Contemporary yarn dyed with modern cudbear

Cudbear powder prepared from British lichens in the mid-20th century by Robert Hill was used to dye woollen yarn from Brush Creek Wool Works (Pennsylvania, USA). Amounts of ingredients were chosen to roughly simulate a Worth recipe for dyeing purple. Yarn samples were dyed with and without acidification of the dye bath (see Appendix 1 for preparation details).

Worth yarn samples

Four yarn samples accompanied by recipes in the Purple section of the book were analysed: WHW 60 (4 lb cudbear and ⅛ pint – 71 mL – grain spirits), WHW 61T (16 lb cudbear and ¾ pint grain spirits), WHW 61M (7 lb cudbear and ¼ pint grain spirits) and WHW 61B (3 lb cudbear and ⅛ pint grain spirits). A Greyish Drab yarn sample, WHW 47L (dyed with 20 lb M. madder, 4 lb Gr. Fustic and ¾ lb cudbear, no dyeing aids) was also analysed.

Chemicals

See Appendix 2 for a list of chemicals and their sources.

Analytical methods

FORS was performed using a Cary 50 spectrophotometer (Varian, now Agilent) with an optical fibre and a Barrelino accessory (Harrick Scientific). Yarn or bunched fibre samples were placed on white, unbuffered blotter or Whatman 1 filter paper for measurements; the same support was also used to collect a baseline spectrum. The Kubelka-Munk transform was applied to smoothed data to obtain the equivalent absorbance spectra. The second derivatives of absorbance curves were taken to distinguish the absorbance maxima of the components of the complex absorbance band. Averages of three to five readings of each sample were used. The wavelength accuracy of the smoothed, corrected absorbance spectra is ± 3 nm.

Fluorescence spectra of small yarn samples were obtained with a Cary Eclipse fluorescence spectrophotometer (Agilent). The samples were mounted in the powder sample holder in the sample compartment. After individual spectra were recorded at selected excitation and emission

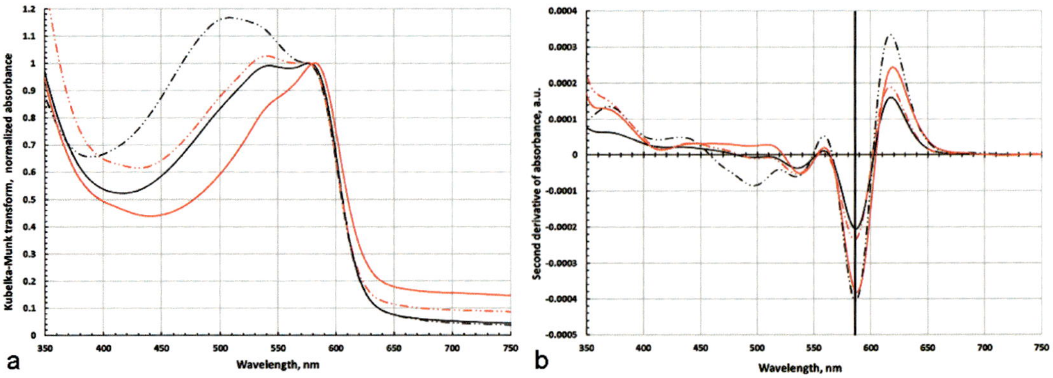

Figure 2 FORS results for reference materials; three to five measurements were averaged for each sample. (a) Normalised Kubelka-Munk transforms of reflectance data for Schweppe yarns dyed with archil extract (black lines), and for contemporary wool dyed with Hill cudbear (red lines); solid lines, alkaline conditions and dashed lines, acidic conditions. (b) Second derivatives of the transforms in (a).

wavelengths, data were collected in the 3D mode, recording emission spectra from 430 to 700 nm for excitation wavelengths at 10 nm intervals from 420 to 640 nm. To avoid fluorescence of the wool substrate, only visible excitation wavelengths were used. The excitation filter was set to Auto and the 430–1100 nm emission filter was used. Monochromator slit widths and the emission photomultiplier voltage were adjusted to optimise emission intensity. Nominal slit widths no greater than 5 nm were used. Savitzky-Golay smoothing was applied to individual traces, and the data plotted as excitation and emission spectra and also as excitation-emission matrices (contour graphs), with fluorescence intensity as the value dimension.

For HPLC–DAD analysis each yarn sample (up to 1 mg) was treated with 300 μL of a mixture of 0.1 M oxalic acid in pyridine/water (1:1, v/v; OAPW in captions) at about 80 °C for one hour. The extracts were centrifuged for 5 min at 12,100 × g, and the supernatants dried in a desiccator under vacuum. The residues were dissolved in 60 μL of methanol:water (1:1, v/v) and the resulting solutions centrifuged for 5–8 min at 12,100 × *g*. Aliquots of these supernatants were injected directly into the HPLC–DAD system.

The HPLC–DAD instrumentation consisted of an 1100 series Agilent Technologies G1322A degasser, G1311A quaternary pump, an auto sampler G1313A and a G1315A diode array detector (DAD). Dye separation was performed on an Agilent Poroshell 120 SB-C18 reverse-phase column (4.6 × 75 mm, 2.7 μm particle size), which was used with a Poroshell SB-C18 pre-column (4.6 × 5 mm, 2.7 μm). The volume of all samples injected was 20 μL. An aliquot of methanol was run as a solvent blank before each sample, and the blank run subtracted from that of each sample.

For the mobile phase, a gradient of 0.1% (v/v) of formic acid in deionised water (A) and 0.1% (v/v) of formic acid in acetonitrile (B) was used. The gradient was 90% A for 3 min, decreasing linearly to 12% A in 40 min, and then held for 7 min. The system was then brought back to 90% A and held for 5 min. The DAD recorded data at five wavelengths: 254 nm, 360 nm, 420 nm, 500 nm and 560 nm.

Experimental results and discussion

Reference materials

The yarns dyed with Hill cudbear have spectral properties in the visible range that are very similar but not identical to the Schweppe yarns dyed with archil extract, as can be seen for the normalised absorbance spectra (Fig. 2a). The acidity of the dye bath affects the colour of the yarn; examples dyed in mildly acidic or basic conditions are shown. The more acidic the bath, the greater the absorbance at *c.*500 and 540 nm compared to that at the longest wavelength maximum between 580 and 590 nm. These data are in close agreement with results in the literature for

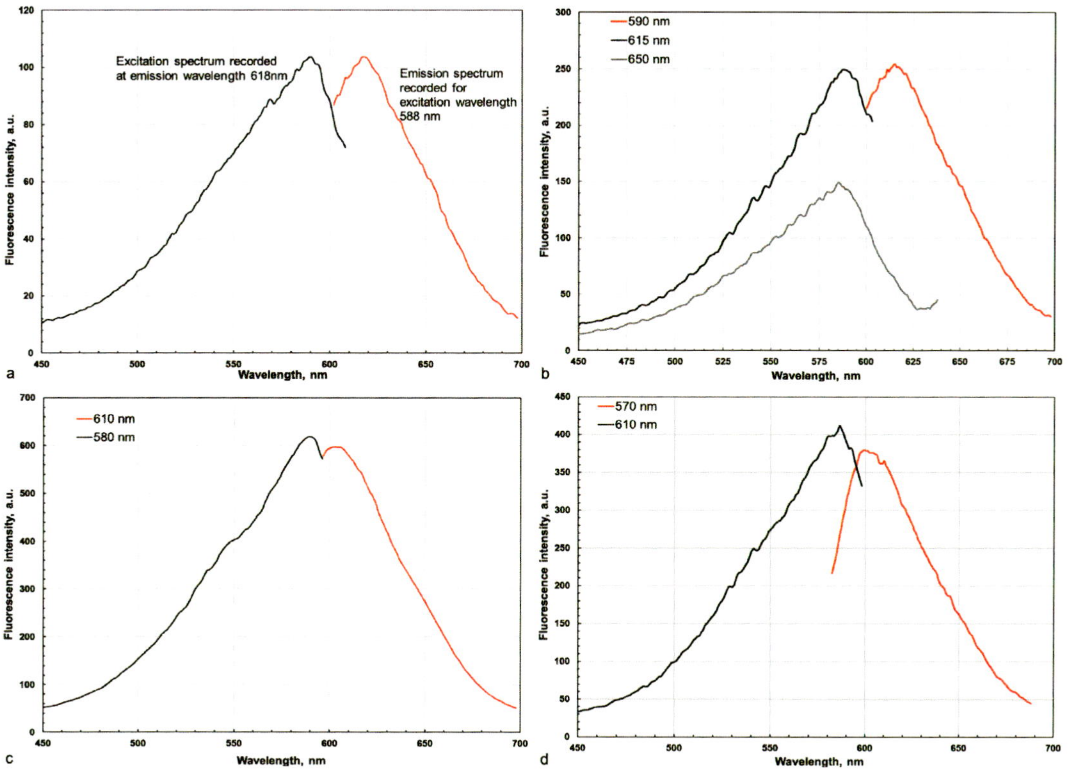

Figure 3 Excitation spectra (black) and emission spectra (red) of reference yarns: (a) Schweppe #300, archil extract plus ammonia; (b) Schweppe #647, archil extract plus acetic acid, alum and tartar; (c) Hill cudbear on wool; (d) Hill cudbear on wool with stannous chloride in nitric acid. Each excitation spectrum was obtained at the emission maximum indicated on the graph, and each emission spectrum was recorded when the sample was excited with monochromatic light at the wavelength indicated. In (b), the grey trace is the excitation spectrum recorded for fluorescence at 650 nm. The fact that this spectrum is the same shape as the excitation spectrum for emission at 610 nm suggests that the same compound is responsible for fluorescence at both wavelengths.

orchil on parchment or silk.[12] When the second derivative of absorbance is plotted, a distinct trough at 585–590 nm is seen in all cases (Fig. 2b). This feature corresponds to the absorbance of the longest wavelength component in the complex absorbance spectrum. The location and shape of this trough do not vary significantly with the conditions of dyeing.

Fluorescence excitation and emission spectra of the reference yarns dyed with archil or cudbear are shown in Figure 3. In all cases, the maximum of the excitation spectrum is at 588 ± 2 nm, which is within the accuracy of the measurements. The maxima of the archil emission spectra have a larger Stokes shift than the cudbear emission spectra. These features appeared to be independent of the dye-bath acidity. They are in general agreement with excitation and emission spectra reported by Clementi and co-authors and Melo *et al.* for orchil on wool and silk,[13] and by Aceto *et al.* and Melo *et al.* for orchil-dyed parchment and illuminated manuscripts, respectively.[14]

The fluorescence intensity contours of the archil and cudbear reference materials are very similar (Fig. 4). In these matrices the fluorescence wavelength is plotted on the X-axis, the excitation wavelength on the Y-axis and the fluorescence intensity is indicated by coloured contours as shown in the legend. Only one region of high intensity fluorescence was observed for all the reference yarns, regardless of the pH and the ratio of dye to wool, and despite the fact that archil and cudbear both consist of several different coloured compounds.[15] The contours are similar in shape and location to those reported by Clarke for cudbear.[16]

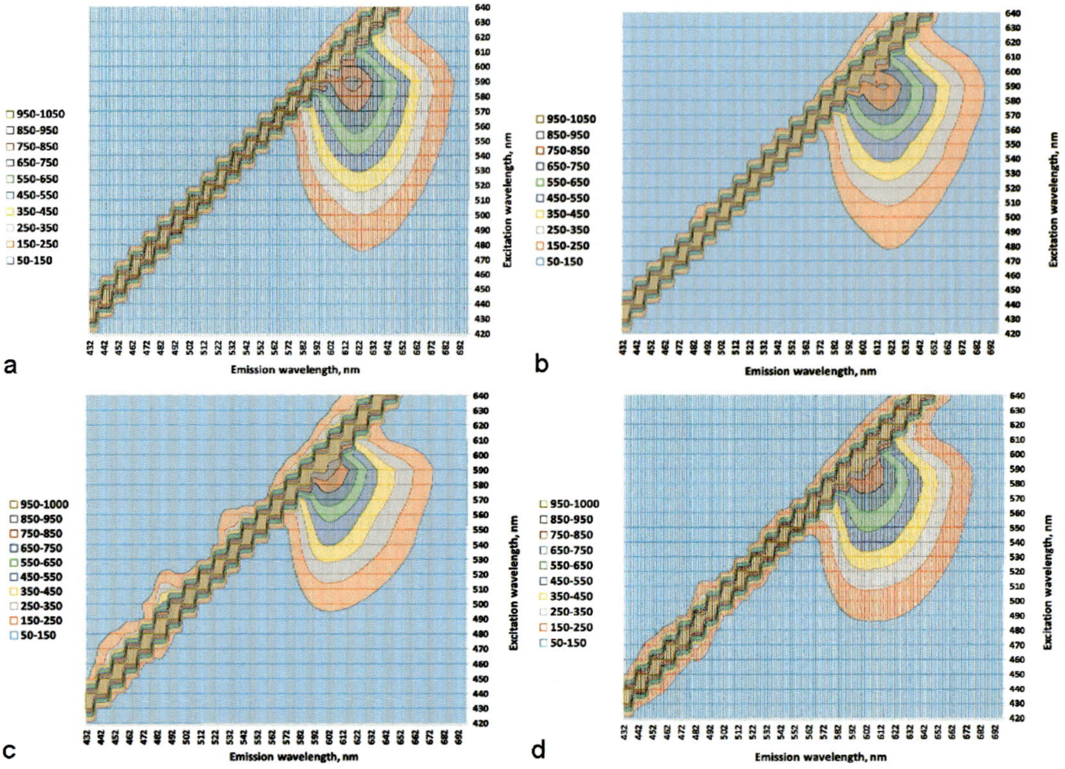

Figure 4 Excitation-emission matrices of fluorescence for reference yarns. The colour scale indicates fluorescence intensity. The diagonal line in each matrix is due to scattered excitation light. (a) Schweppe #300; (b) Schweppe #647; (c) Hill cudbear; (d) Hill cudbear with stannous chloride in nitric acid.

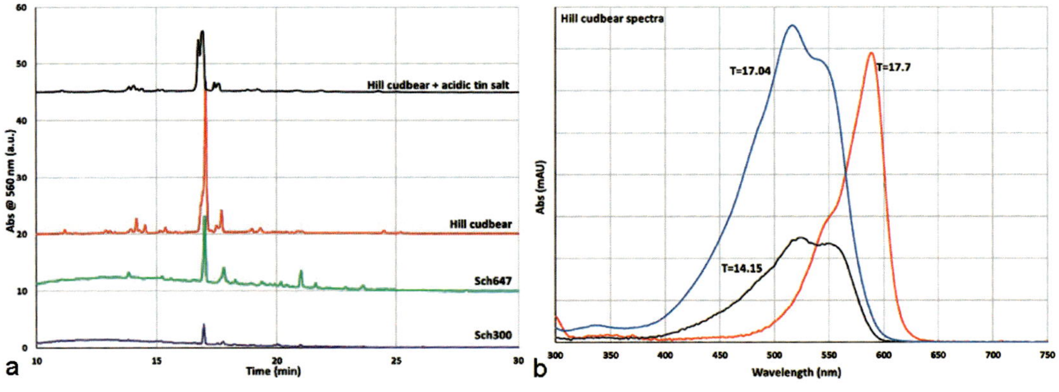

Figure 5 (a) HPLC chromatograms recorded at 560 nm for the OAPW extracts from reference materials; samples as indicated on the traces; (b) DAD spectra of peaks eluting at the times indicated for the extract from the yarn dyed with Hill cudbear.

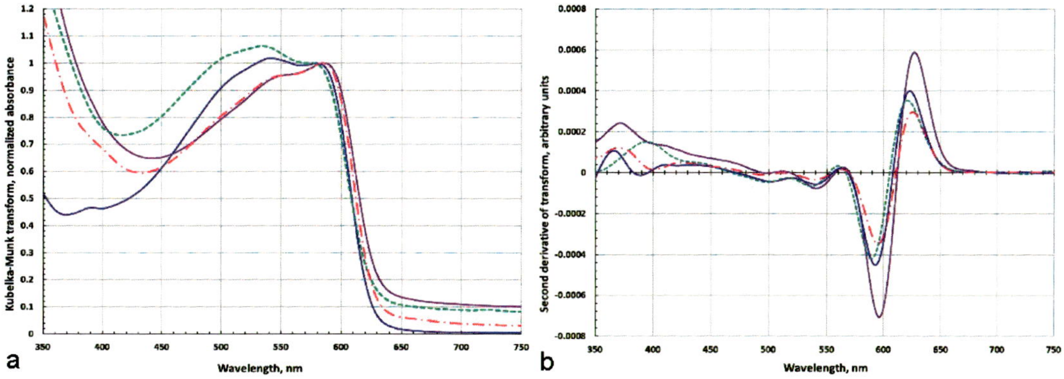

Figure 6 (a) Normalised Kubelka-Munk transforms of FORS data for Worth purple yarns; solid purple line WHW60, dash/dot red line, WHW61B, solid blue line WHW61T, dotted green line, WHW 61M. (b) Second derivatives of the absorbance curves in (a).

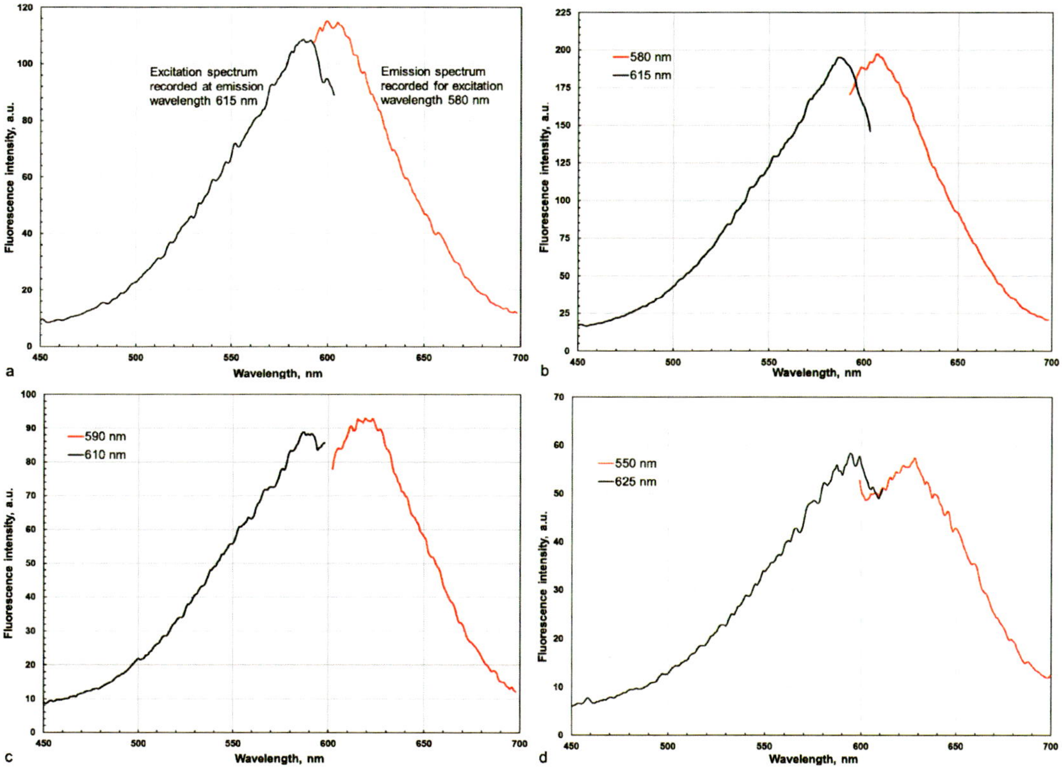

Figure 7 Excitation spectra (black) and emission spectra (red) of Worth purple samples dyed with cudbear only: (a) WHW 60; (b) WHW 61B; (c) WHW 61M; (d) WHW 61T. Each excitation spectrum was obtained at the emission maximum indicated on the graph, and each emission spectrum was recorded when the sample was excited with monochromatic light at the wavelength indicated.

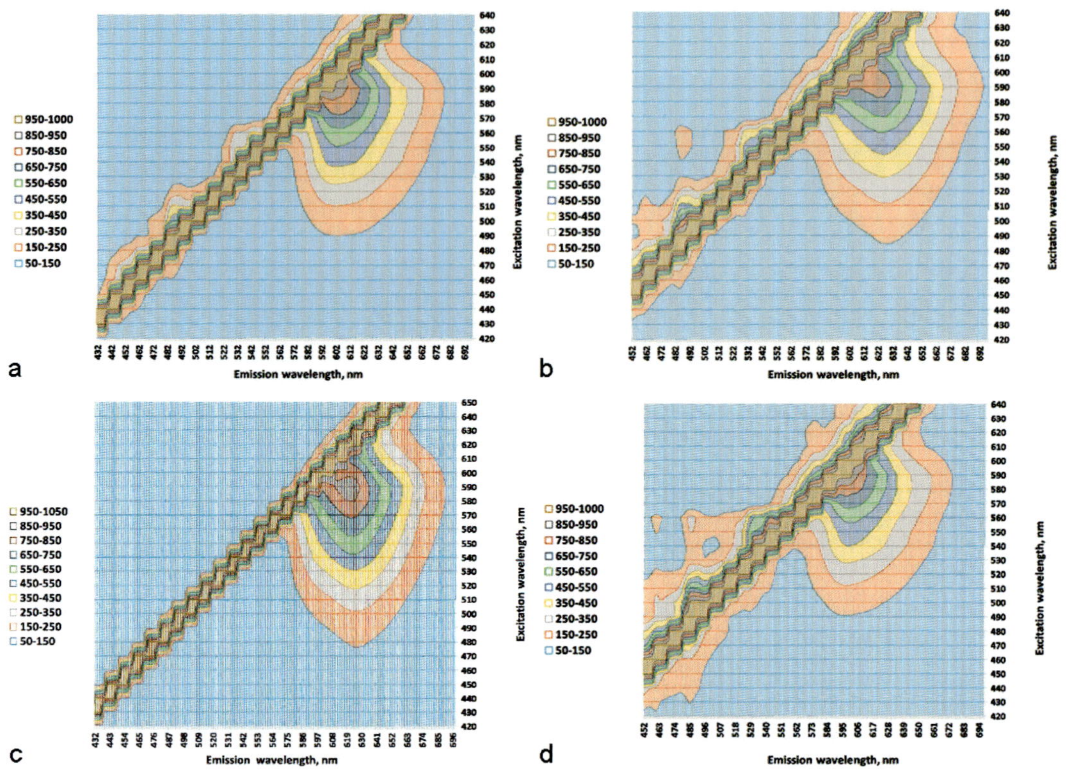

Figure 8 Excitation-emission matrices of fluorescence for Worth purple samples: (a) WHW 60; (b) WHW 61T; (c) WHW 61M; (d) WHW 61B.

Reports of HPLC analyses of archil (orchil) have recently appeared in the literature, but there is very limited data on cudbear.[17] In the current study, HPLC analyses of extracts from the cudbear- and archil-dyed reference yarns indicated that these dyestuffs contain some similar components, but in different amounts (Fig. 5). Using a detector wavelength of 560 nm enhanced detection of some components by the DAD. The chromatograms (Fig. 5a) and the spectra of the coloured compounds eluting between 17 and 18 min (Fig. 5b) are similar for all the reference materials. In addition, the cudbear samples have a characteristic band at about 14 min (Fig. 5b), absent from the Schweppe archil references. Taken together with the FORS and fluorescence results, these characteristics may serve as a positive identification of cudbear, even though the individual compounds in the dyestuffs cannot be identified chemically using these results. It should be noted that the spectra of these HPLC bands closely resemble absorbance spectra reported under certain conditions for resazurin and resorufin, two phenoazine compounds that are similar to components of orcein.[18]

Worth samples

The normalised absorbance curves for the Worth purple samples are shown in Figure 6a. Both the curve shapes and the effect of acid in the dye bath are very similar to those for the reference materials (see Fig. 2a). The shorter wavelength bands of the two Worth samples with the least acidic tin salt in the recipe have relatively lower absorbance. The prominent feature in the second derivative curves is the long wavelength trough, regardless of the conditions during dyeing (Fig. 6b). The fluorescent excitation maxima are at about 588 nm, and the emission spectra are similar in shape and position to those of the Hill cudbear samples (Fig. 7). The fluorescence contours for these Worth purple samples display

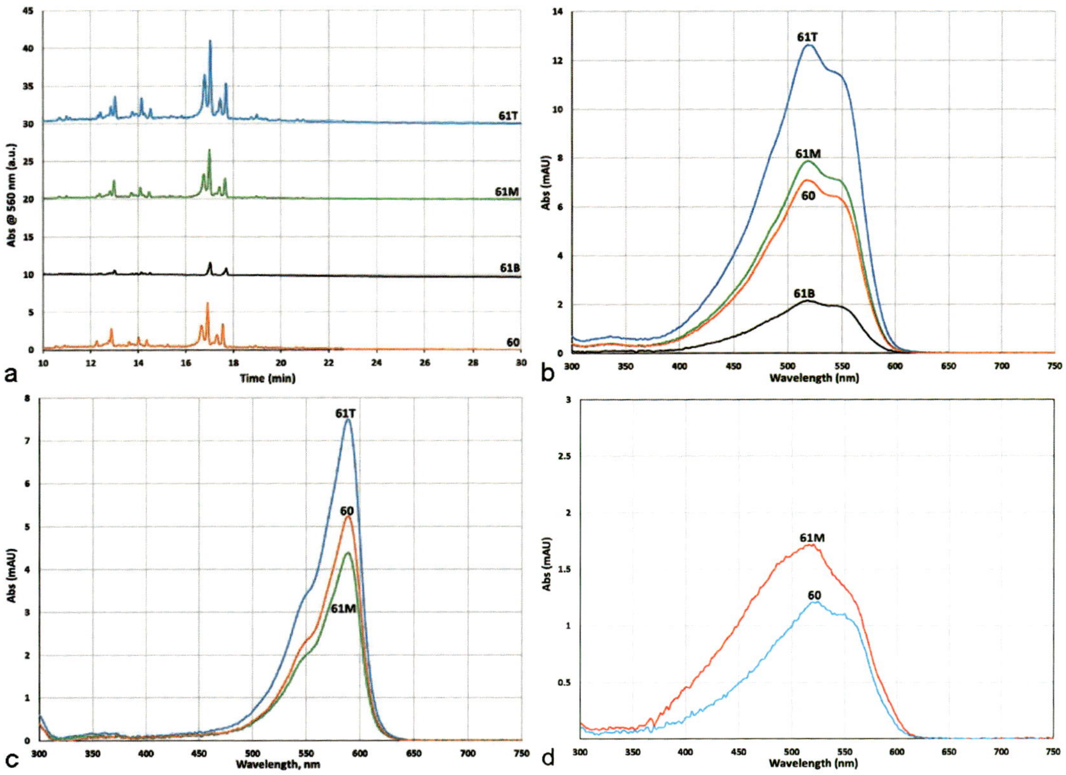

Figure 9 (a) HPLC chromatograms recorded at 560 nm of OAPW extracts from Worth purple samples as indicated on the traces; (b) DAD spectra of the bands eluting at 17.0 min; (c) DAD spectra of the bands eluting at 17.6 min; (d) DAD spectra of the bands eluting at about 14 min.

the same shape and features as the reference materials (compare Figs 8 and 4). The spectroscopic data for Worth's purple yarns are all consistent with the presence of cudbear and indicate that, at least qualitatively, he did follow the recipes accompanying his purple yarns.

Further evidence for Worth's use of cudbear in his recipes for purple was obtained by HPLC of extracts from these samples (Fig. 9a). Major peaks between 17 and 18 min were detected for all four extracts, as well as the smaller band at about 14 min. The spectra of the major peaks have the same characteristics as those of the extracts of yarns dyed with reference materials (compare Fig. 9b,c,d to Fig. 5b), further supporting the presence of cudbear. This deduction can be made despite the lack of chemical identification of the dye compounds present.

The possibility of detecting cudbear when it is a minor dye component on a yarn was tested on a Greyish Drab sample (WHW47L) in Worth's book (Fig. 10a). The recipe accompanying this sample indicates that cudbear should be contributing only 3.3% of the total weight of dye, and be present at just over 1% of the weight of wool. The absorbance spectrum of WHW47L is shown in Figure 10b. Although this spectrum is not useful for identifying the dyes present, the negative trough at about 590 nm in the second derivative curve is distinct, and suggests that a lichen dyestuff may be present (Fig. 10c). This observation is supported by the fluorescence excitation and emission curves (Fig. 10d) and the fluorescence contour, in which the characteristic pattern of cudbear fluorescence is evident, despite the background of low level fluorescence at shorter excitation and emission wavelengths (Fig. 10e). The featureless, low intensity fluorescence in this latter region of the matrix might be due to the large amounts of madder and fustic in the sample compared to the cudbear content.

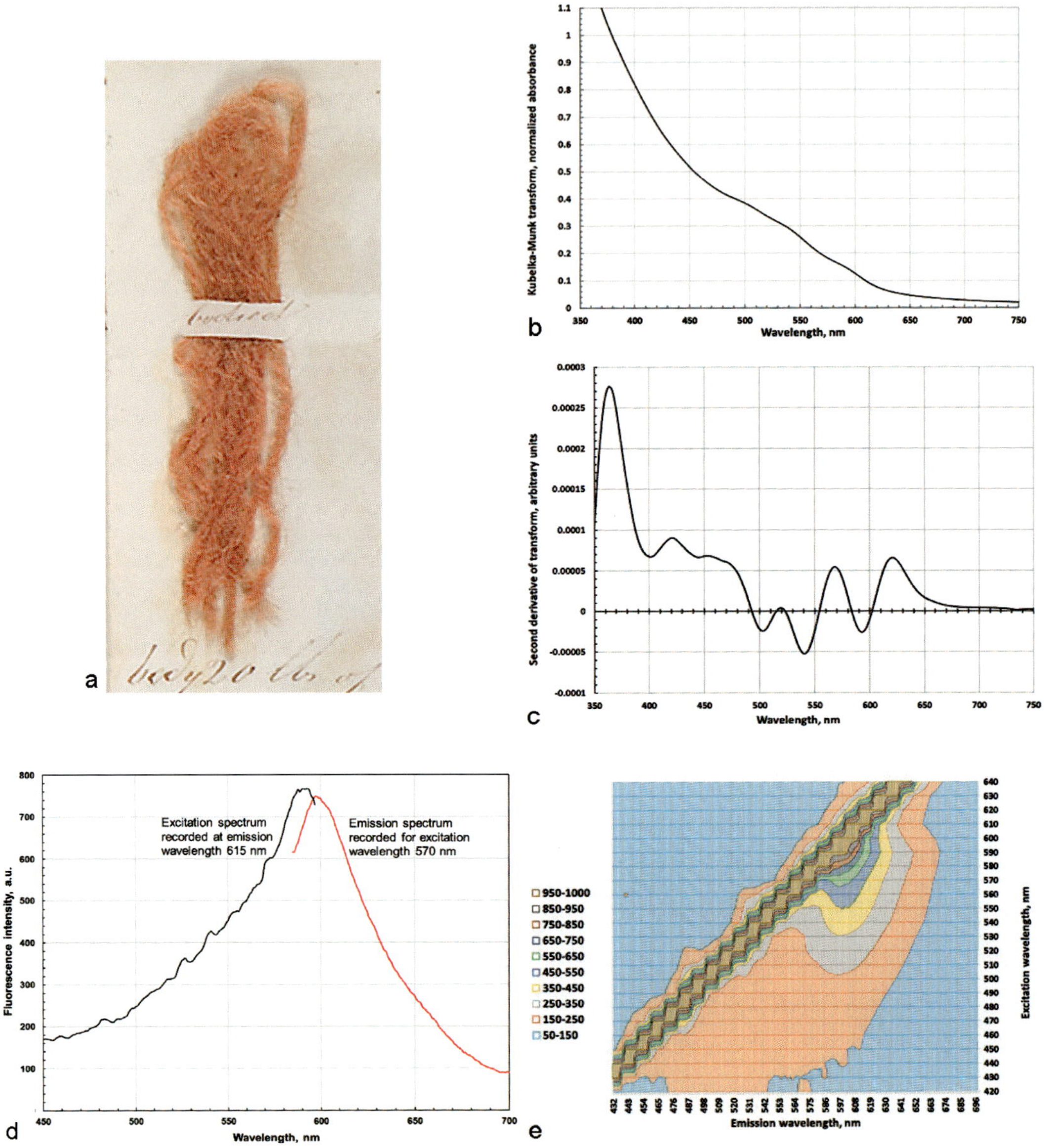

Figure 10 (a) Worth Greyish Drab yarn sample WHW 47L in the Worth book. Gift of Lloyd and Margit Cotsen (photo © Museum Associates / LACMA Conservation Center by Yosi Pozeilov); (b) Kubelka-Munk transform of reflectance data; (c) second derivative of the transform; (d) fluorescence excitation (black) and emission spectra (red); (e) excitation-emission matrix of fluorescence.

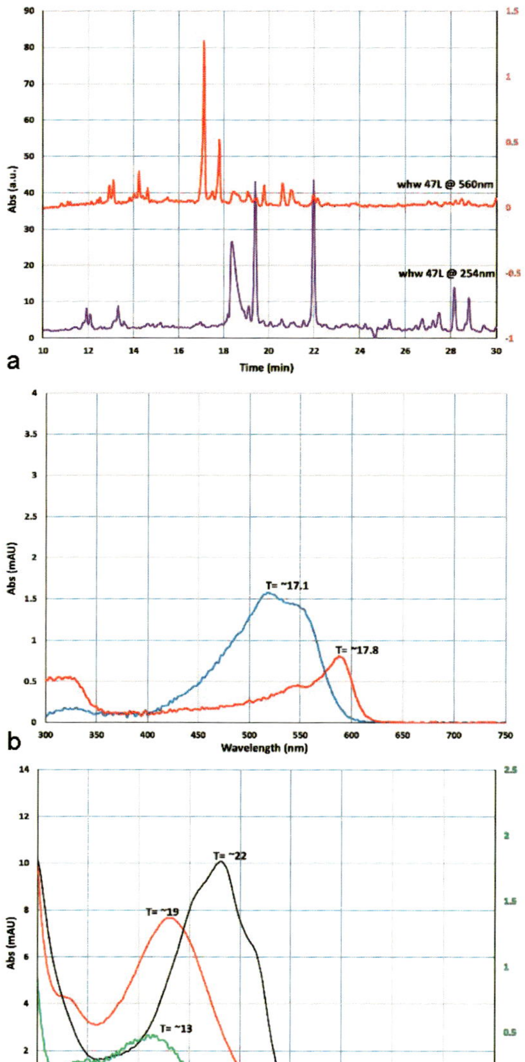

Figure 11 (a) HPLC chromatograms of OAPW extract from Worth sample 47L, recorded at 254 and at 560 nm as indicated on the traces; (b) DAD spectra of bands detected at 560 nm eluting at 17.1 and 17.8 min and ascribed to the small amount of cudbear present; (c) DAD spectra of the bands detected at 254 nm, eluting at approximate times of 13, 19 and 22 min and ascribed to madder components.

HPLC chromatograms obtained at 254 and 560 nm for the extract from Worth sample 47L are shown in Figure 11a. The spectra of the coloured peaks in the 560 nm chromatogram are presented in Figure 11b, and those of the coloured peaks in the 254 nm chromatogram are shown in Figure 11c. The chromatogram recorded at 360 nm had a peak at 13.3 min that corresponds to a component of fustic dye (data not shown). Each chromatogram in Figure 11 highlights the presence of a dyestuff in the recipe: madder components in the 254 nm trace, and cudbear components chromatographing between 17 and 18 min in the 560 nm trace. The detection of cudbear agrees with the FORS and fluorescence results, and confirms the suggestion that cudbear is present in this drab yarn. Notably, evidence for cudbear was obtained by the spectroscopic methods in the presence of a large excess of other dyes in this Worth sample.

Conclusion

The many details in Worth's book provide valuable information on large-scale early Victorian dyeing practice. The recipes corroborate the literature suggesting that cudbear was used extensively in carpets in the mid-19th century. For the samples investigated, the experimental evidence implies that Worth's recipes were followed – at least qualitatively – to create those samples.

Concomitant application of several analytical methods to confirm the presence of orcein in historic materials has recently been demonstrated.[19] In the current investigation, the three analytical methods used (FORS, fluorescence spectrophotometry, HPLC–DAD) provided data that confirmed the presence of cudbear in the Worth samples. FORS data were suggestive of its presence. Because cudbear is highly fluorescent on woollen yarn, fluorescence spectroscopy is a highly sensitive means of detecting the dyestuff.[20] In this work, the method provided evidence of cudbear in the presence of an excess of other dyes. HPLC–DAD analyses of the Worth yarn samples were supportive of the non-destructive spectroscopic results that suggested cudbear was present. These methods taken together allowed detection of cudbear despite the lack of data identifying the chemical components present in the dyestuff.

Even with this lack of chemical information, the results of the investigation suggested that the cudbear Worth used resembles the modern British cudbear more closely than it does the European archil. The data also support the consideration of archil and cudbear as a group of dyestuffs in the same way as the various madders are grouped together. The numerous plant sources for dyestuffs in each of these groups will yield several coloured compounds common to both, but in varying relative amounts. The quantities of the components detected on a fibre will depend on the species of plant, the growth location and conditions, the preparation of the dyestuff and the dyeing process. However, for both the archil/cudbear and the madder groups, detection of particular 'signature components' can signal the presence of the dyestuff whether or not the chemical identities of the components are known.

Appendix 1

Preparation of cudbear-dyed yarn

Having added 33 mg dye powder and 1.1 mg recrystallised hydrated aluminium potassium sulfate ($KAl(SO_4)_2{\cdot}12H_2O$) to 125 mL of MilliQ water, the dye bath was heated to ~75 °C and stirred for several hours. Not all of the dye went into solution. The suspension was cooled to room temperature and left in the dark overnight. The next day the bath was reheated and water added to 200 ml. No additional dye appeared to go into solution. The suspension was filtered through unsized polyester cloth (TestFabrics), and reheated. Approximately 0.5 g of undyed woollen yarn was added and the dye bath was kept at ~75 °C, with stirring, for about 6 hours. The heat was turned off, and the yarn left in the bath in the dark overnight. The next day the bath was reheated and stirred for an additional 5 hours. The bath was allowed to cool and the yarn – now a light lavender – was removed.

Both the yarn and the dye bath were divided into two. To one half of the bath containing one half of the yarn, was added 50 µL of a solution prepared by adding 14.5 mg hydrated stannous – (tin(II) – chloride to 40 mL of ~38% nitric acid. This acidic tin-salt solution was used to simulate the 'grain spirits' used by Worth, and lowered the pH of the dye bath to about 3. The bath was stirred for 20 min and the yarn, now distinctly reddish, was removed and dried at room temperature. To the other half of the dye bath was added 0.1 mL aqueous ammonia solution (27–30%, v/v), which raised the pH to about 9, and the other half of the light lavender yarn was added. After 30 minutes of stirring, the yarn was a slightly deeper lavender. It was removed and allowed to dry.

Appendix 2

Chemicals and suppliers

- Acetonitrile (HPLC grade), methanol (HPLC grade), water (HPLC grade), oxalic acid (lab grade) and 0.1% (v/v) formic acid in acetonitrile (Optima LC/MS): Fisher Scientific.
- Formic acid (LC–MS grade) and pyridine (LC–MS grade): Thermo Scientific.
- Drierite with and without indicator: Acros Organics.
- Denatured anhydrous ethanol (90% ethanol, 5% methanol, and 5% isopropanol, HPLC Grade): Alfa Aesar.
- Concentrated nitric acid: Sigma Aldrich.
- Water was purified by a Millipore Simpak MilliQ system until it reached a resistivity of 18.2 Mohm. Ammonia solution (27–30%, v/v), aluminium potassium sulfate (recrystallised from purified water in the laboratory) and stannous chloride were purchased from local suppliers.

Editor's note and authors' addresses

*Author for correspondence: terrys@lacma.org.
Paper received 13 March 2018; revised version received 16 September 2018. This paper was presented at *Dyes in History and Archaeology* 36, Hampton Court, UK, 26–28 October 2017.

[1] Conservation Center, Los Angeles County Museum of Art (LACMA), 5905 Wilshire Blvd, Los Angeles, CA 90036, USA

Acknowledgements

The authors thank Dr David Hill for the generous gift of cudbear prepared by Dr Robert Hill; Lloyd and Margit

Cotsen for the donation of the Worth sample book to the Los Angeles County Museum of Art; Dr Tom Learner and Art Kaplan of the Getty Conservation Institute for samples of Schweppe reference yarns; Dr Victor Chen of the Indiana Museum of Art for advice on HPLC; and Yosi Pozeilov and Dr Charlotte Eng of Los Angeles County Museum of Art, for expert photography and many helpful discussions, respectively.

Notes and references

1. Habib, V. (2014) 'The colours of cudbear: continuity and change in wool and linen dyeing in eighteenth century Scotland', *Review of Scottish Culture* 26, pp. 65–86.
2. Whitworth, I. and Koren, Z.C. (2016) 'Orchil and Tyrian purple: two centuries of Bedfords from Leeds', *Ambix* 63, 3, pp. 244–67; Habib 2014 (see note 1 above).
3. Cardon, D. (2007) *Natural Dyes: Sources, Tradition, Technology and Science*, London, pp. 510–11; Schweppe, H. (1993) *Handbuch der Naturfarbstoffe*, Hamburg, Nikol Verlagsgesellschaft mbH & Co. KG, pp. 528–9; Casselman, K. Diadick (2001) *Lichen Dyes: The New Source Book*, New York, Dover Publications, Inc.
4. Hofenk de Graaff, J.H. (2004) *The Colourful Past: Origins, Chemistry and Identification of Natural Dyestuffs*, with contributions from W.G.Th. Roelofs and M.R. van Bommel, London and Riggisberg, pp. 274–8.
5. Love, T. (1854) *The Art of Cleaning, Dyeing, Scouring, and Finishing, on the most approved English and French Methods* ..., London, Longman, Brown, Green, and Longmans; Molony, C. (1837) *Molony's Masterpiece on Wool, Silk and Cotton Dyeing: Containing his Best Receipts, without the least Reserve; according to his Practice in Great Britain and America*, Lowell: printed by Dearborn & Bellows; Celoria, F. (1971) 'Archaeology and dyestuffs: observation centring on some 19th-century evidence of dyestuffs and equipment used in England before the development of artificial dyes; with tables, comments and a general bibliography', *Science and Archaeology* 6 (Stafford, George Street Press), pp. 15–38.
6. Kok, A. (1966) 'A short history of the orchil dyes', *The Lichenologist* 3, 2, pp. 248–72.
7. Thompson, M. (2002) *Woven in Kidderminster: An Illustrated History of the Carpet Industry in the Kidderminster Area including Stourport, Bridgnorth and Bewdley 1735–2000*, Kidderminster, David Voice Associates, p. 36.
8. Schaeffer, T.T. (2017) 'Mr. Worth's woollen yarns: a Victorian book of dyed samples with recipes', in J. Kirby (ed.), *The Diversity of Dyes in History and Archaeology*, London, pp. 304–16.
9. Love 1854 (see note 5 above); Molony 1837 (see note 5 above); Celoria 1971 (see note 5 above).
10. There is no recipe for grain spirits in the sample book. In dyers' manuals roughly contemporary with the Worth sample book this term is applied to acidic solutions of tin salts; e.g. Love, T. 1854 (see note 5). 1 ounce (oz) is equivalent to 28.35 g, 1 pound (lb, avoirdupois = 16 oz) is equivalent to 453.6 g; 1 pint (UK imperial measure) is equivalent to 0.568 L. See Young, R.A. (1996) *Measure for Measure*, Madera CA, Sequoia Publishing for conversions into metric units.
11. Schweppe 1993 (see note 3 above), p. 93.
12. Aceto, M., Arrais, A., Marsano, F., Agostino, A., Fenoglio, G., Idone, A. and Gulmini, M. (2015) 'A diagnostic study on *folium* and *orchil* dyes with non-invasive and micro-destructive methods', *Spectrochimica Acta Part A: Molecular and Biomolecular Spectroscopy* 142, pp. 159–68; Clementi, C., Miliani, C., Romani, A. and Favaro, G. (2006), 'In-situ fluorimetry: a powerful non-invasive diagnostic technique for natural dyes used in artefacts: Part I. Spectral characterization of orcein in solution, on silk and wool laboratory-standards and a fragment of Renaissance tapestry', *Spectrochimica Acta Part A* 64, 4, pp. 906–12; Melo, M.J., Nabais, P., Guimarães, M., Araújo, R., Castro, R., Oliveira, M.C. and Whitworth, I. (2016) 'Organic dyes in illuminated manuscripts: a unique cultural and historic record', *Philosophical Transactions of the Royal Society A* 374, doi: 10/1098/rsta.2016.0050.
13. Clementi *et al.* 2006 (see note 12 above); Melo *et al.* 2016 (see note 12 above).
14. Aceto 2015 (see note 12 above); Melo 2016 (see note 12 above).
15. Beecken, H., Gottschalk, E.-M., Gizycki, U. v., Krämer, H., Maassen, D., Matthies, H.-G., Musso, H., Rathjen, C. and Zdhorszky, Ui. (2003) 'Orcein and litmus', *Biotechnic & Histochemistry* 78, 6, pp. 289–302; Schweppe 1993 (see note 3 above), pp. 517–18; Cardon 2007 (see note 3 above), pp. 487–8; Doherty, B., Gabrieli, F., Clementi, C., Cardon, D., Sgamelotti, A., Brunetti, B. and Miliani, C. (2014) 'Surface enhanced Raman spectroscopic investigation of orchil dyed wool from *Rocella tinctoria* and *Lasallia pustulata*', *Journal of Raman Spectroscopy* 45, pp. 723–9.
16. Clarke, M. (1999) 'A new technique for the non-destructive identification of organic pigments, dyes and inks in-situ on early medieval manuscripts, using 3-D fluorescence reflectance spectroscopy', in *Art '99: 6th International Conference on Non-Destructive Testing and Microanalysis for the Diagnostics and Conservation of the Cultural and Environmental Heritage, Rome, May 17th–20th 1999*, organised by the Istituto centrale per il restauro, Associazione italiana prove non-distruttive (AIPnD), 3 vols, Rome, vol. 2, pp. 1421–36.

17. Melo *et al.* 2016 (see note 12 above); Witkowski, B., Ganeczko, M., Hryszko, H., Stachurska, M., Gierczak, T. and Biesaga, M. (2017) 'Identification of orcein and selected natural dyes in 14th and 15th century liturgical paraments with high-performance liquid chromatography coupled to the electrospray ionization tandem mass spectrometry (HPLC–ESI/MS/MS)', *Microchemical Journal* 133, pp. 370–79; Karmous, T. and Ayed, N. (1998) 'Orchil lichens of Tunisia: chromatographic identification and dyeing system', *Dyes in History and Archaeology* 16/17, pp. 119–28; Karmous, T., Ayed, N. and Nowik, W. (1997) 'Caractérisation par chromatographie des lichens tinctoriaux de Tunisie', *Analusis* 25, 9–10, pp. 321–9; Hofenk de Graaff 2004 (see note 4 above), pp. 278–9.
18. Bueno, C., Villegas, M.L., Bertolotti, S.G., Previtali, C.M., Neumann, M.G. and Encinas, M.V. (2002) 'The excited-state interaction of resazurin and resorufin with amines in aqueous solutions: photophysics and photochemical reaction', *Photochemistry and Photobiology* 76, 4, pp. 385–90; Porcal, G.V, Altamirano, M.S., Glusko, C.A., Bertolotti, S.G. and Previtali, C.M. (2011) 'Photophysical properties of the phenoxazin dyes resazurin and resorufin in soybean lecithin microemulsions', *Dyes and Pigments* 88, 3, pp. 240–46.
19. Melo *et al.* 2016 (see note 12 above).
20. Clementi, C., Miliani, C., Romani, A., Santmaria, U., Morresi, F., Mlynarska, K. and Favaro, G. (2009) 'In-situ fluorimetry: a powerful non-invasive diagnostic technique for natural dyes used in artefacts. Part II. Identification of orcein and indigo in Renaissance tapestries', *Spectrochimica Acta Part A* 71, pp. 2057–62; Clementi *et al.* 2006 (see note 12 above); Melo *et al.* 2016 (see note 12 above).

A Non-Destructive Approach to Dye Analysis for Typical 17th–19th-Century East Asian Textiles

Yoshiko Sasaki[1]* and Ken Sasaki[2]

Abstract Spectroscopy is important for the non-destructive characterisation of natural dyestuffs used for precious historical textiles. On-site analytical methods are particularly important for historical Japanese kimonos in museums or temples. Compact and portable spectrometers using optical fibres are powerful devices for on-site analysis, although their performance is limited compared to that of benchtop instruments. We have developed a portable system to measure visible reflectance and multi-excited fluorescence spectra and applied it for on-site dye analysis of 17th–18th-century Japanese kimonos and their fragments. The efficiency of this system was verified through comparison with the results of benchtop instruments. It was found that information on minor components of multiple dyes was easily obtained by combining visible reflectance and fluorescence emission spectra with multiple excitation wavelengths. This paper reports on the efficiency and limitations of our on-site analytical method and introduces historical dyeing of Japanese kimonos.

Introduction

Spectroscopy is important for the non-destructive and microsampling characterisation of natural dyestuffs used in precious historical textiles. We have developed methodologies for both approaches to analyse natural and synthetic dyes in various textile materials.[1] Our analytical protocols fall into three categories:[2]

- non-destructive on-site analysis using portable small spectroscopic devices;
- non-destructive analysis using benchtop instruments in a laboratory when the movement of a textile from its storage site is permitted;
- microsampling analysis using various analytical instruments in a laboratory.

The movement of precious national treasures from their storage sites to laboratories for analysis with benchtop instruments is generally restricted, therefore on-site analysis in museums is important, especially for highly valuable artefacts. The recent development of handheld optical devices enables scientific evidence for dyes to be acquired readily by on-site measurement.[3] However, when sampling of a textile is permitted, more precise analysis can be performed using benchtop instruments, for example by excitation-emission matrix correlation fluorescence three-dimensional spectroscopy (fluorescence EEM 3D),[4] micro-region spectroscopy and integrating sphere–attached ultraviolet-visible (UV-VIS) spectroscopy.[5]

Generally, combinations of several analytical techniques increase the accuracy of identification. We have already developed a non-destructive system combining visible reflectance with fluorescence which has been applied successfully in the dye analysis of 17th–18th-century Japanese kimono fragments.[6] This paper describes the

Table 1 Spectroscopic data of visible reflection and fluorescence spectra for typical Japanese dyestuffs used in the Edo era (17th–19th century).

Colour	Dye (mordant)	Fluorescence		Reflectance	Major detectable chemical component
		Intensity λex (365)/ λex (465)[a]	Counter peaks of EEM spectra[b] λex / λem (nm)	Maximum absorption (nm)	
Red	safflower	+/+++	385, 550/590	530	carthamin
	sappanwood (alum)	−/±	(530/600)	530	brazilein
	Japanese madder (alum)	−/++	550/580	500, 480(sh), 550(sh)	purpurin
	cochineal (alum)	−/±	(530/600)	525, 500(sh), 570(sh)	carminic acid
	lac (alum)	−/±	(530/600)	525, 500(sh), 570(sh)	laccaic acids
Yellow	amur cork tree	+++/+++	360, 460/530	370, 440	berberine
	turmeric	+/+++	460/530	450	curcumin
	eulalia (alum)	−/+	(430/510)	450, 580(sh)	luteolin
	pagoda tree (alum)	−/+	(430/510)	450	quercetin
	gardenia	−/−	–	450, 475(sh)	crocetin
Purple	gromwell (none)	−/−	–	530, 580(sh)	shikonin
	gromwell (alum)	−/−	–	550, 600	shikonin
Blue	indigo	−/−	–	650	indigo

[a] Fluorescence intensity, excitation at 365 nm/465 nm: +++: very strong; ++: strong; +: moderate; ±: weak; −: not detected.
[b] The values in parentheses indicates peak maxima that are very weak in intensity.

efficiency of this optical fibre-assisted system for non-destructive spectroscopic analysis of dyes in 17th-century traditional Japanese and Chinese textile fragments in the collections of the Kyoto Institute of Technology, Japan.

Experimental

Fluorescence spectra were recorded using a Hitachi F-4500 spectrofluorometer with an optical fibre attachment. An Ocean Optics USB 4000 fibre-optic multichannel spectrometer was used as a portable device for the measurement of visible reflectance and fluorescence spectra. The excitation light source for the fluorescence used LED lasers with maximum wavelengths of 365 nm and 465 nm. Dyed reference samples were prepared according to our previous report.[7]

Results and discussion

Spectroscopic overview of traditional East Asian dyes

Visible and fluorescence spectra of dyes for 17th–19th-century Japanese and Chinese silk clothes were measured using handheld and benchtop instruments for optical fibre-derived spectroscopy.[8] Table 1 summarises the characteristic spectroscopic data, including the peak positions of the fluorescence EEM 3D spectra and second derivatives of visible reflectance spectra for the typical ancient Japanese and Chinese dyestuffs described here. The table also summarises the relative intensities and maximum emissions of the fluorescence spectra resulting from excitation wavelengths of 365 nm and 465 nm, which were obtained using a small LED light source effective in portable spectroscopic devices.

The visible reflectance and fluorescence spectra of typical red dyes were as follows. Safflower (*Carthamus tinctorius* L.) exhibited maximum absorption around 530 nm and a strong fluorescence emission at 590 nm after excitation at 465 nm. Sappanwood (*Caesalpinia sappan* L., now *Biancaea sappan* (L.) Tod.) with an alum mordant exhibited a similar maximum absorption around 540 nm. However, this dye could be readily discriminated from safflower by its characteristic non-fluorescence and by differences in their second-derivative peak positions (530 nm in safflower, 550 nm in sappanwood).

In the case of the anthraquinone red dyes, lac (*Kerria lacca* (Kerr, 1782)) and cochineal (*Dactylopius coccus* Costa, 1829, generally used after the 18th century) possessed characteristic

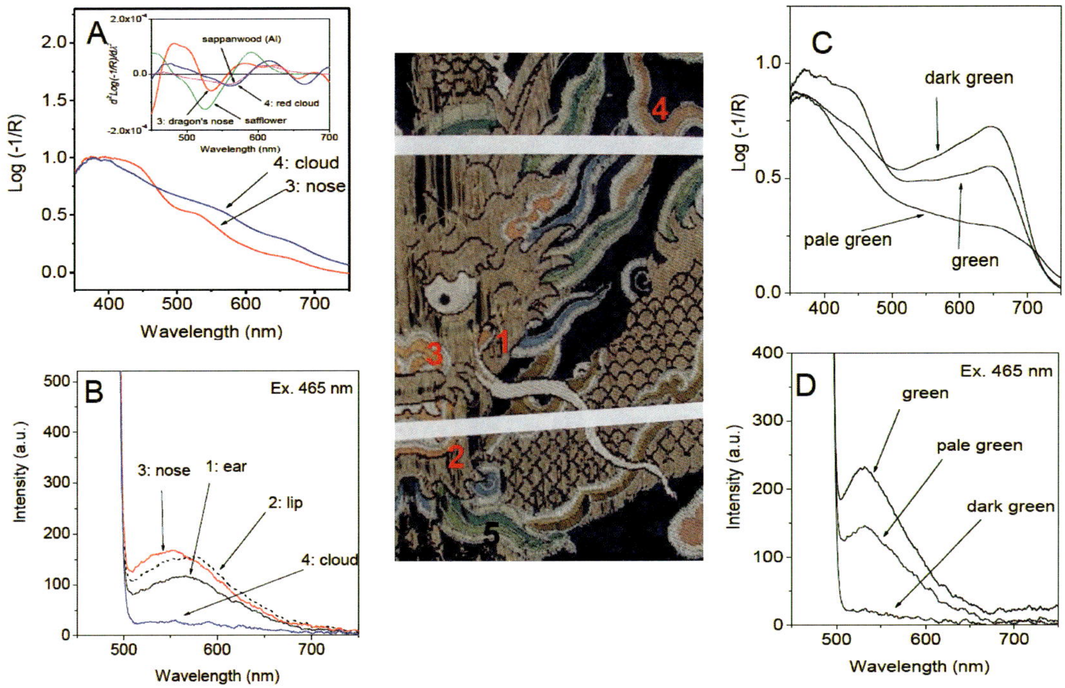

Figure 1 *Satin Fragment with Design of Dragon, Cloud, and Wave on Blue Ground*, early-to-mid Qing dynasty (KIT Collection A-c-8). Visible reflectance and fluorescence spectra of red dyes for the dragon's face and waving cloud, and for green *ungen* in the dragon's beard. A: visible reflectance spectra of red dyes, inset: second derivatives for the dragon's nose 3 and cloud 4; B: fluorescence spectra of red dyes excited at 465 nm; C: visible reflectance spectra of green *ungen*; D: fluorescence spectra of green *ungen* excited at 465 nm.

polynomial visible spectra and weak fluorescence emissions, while Japanese madder (*Rubia akane* Nakai, now *Rubia argyi* (H.Lév. & Vaniot) H.Hara) showed emissions of moderate intensity due to its major purpurin component. Second derivatives of their visible spectra showed characteristic peaks that could easily distinguish anthraquinone dyes from other red dyes such as safflower and sappanwood.[9] This approach was also effective for discriminating Japanese madder (including purpurin) from insect dyes such as lac and cochineal, based on the differences in their peak positions.

The visible reflection spectra for the yellow dyes, amur cork tree (*Phellodendron amurense* Rupr. and *P. chinense* C.K.Schneid.), turmeric (*Curcuma longa* L.), eulalia (*Miscanthus tinctorius* (Steud.) Hack.), pagoda tree (*Sophora japonica* L., a synonym of *Styphnolobium japonicum* (L.) Schott), and gardenia (*Gardenia jasminoides* J.Ellis), had absorption maxima around 450–470 nm. Strong fluorescence emissions around 510–530 nm were observed for amur cork tree and turmeric. The absorption spectra of amur cork tree also showed two peaks, at 360 nm and 460 nm, characteristic of protoberberine alkaloids. Fluorescence intensities in the multi-excitation spectra of 365 nm and 465 nm usefully distinguished amur cork tree from turmeric. These dyes showed almost identical fluorescence emission spectra,[10] but a second peak at $\lambda_{ex}/\lambda_{em}$ = 365 nm/530 nm was specific for the amur cork tree. Thus, the fluorescence intensities excited at 365 nm and 465 nm were effective for discriminating between these dyes using a small portable device. The fluorescence spectra of eulalia and pagoda tree showed very weak emissions with their detection limited by the concomitant emissions of aged silk. No fluorescence was observed for gardenia, which had characteristic polynomial peaks revealed by the second derivative of its spectrum.

The blue dye component indigotin, derived from natural indigo (*Polygonum tinctorium* W.T.Aiton, synonymous with *Persicaria tinctoria* (Aiton) Spach), showed an absorption maximum

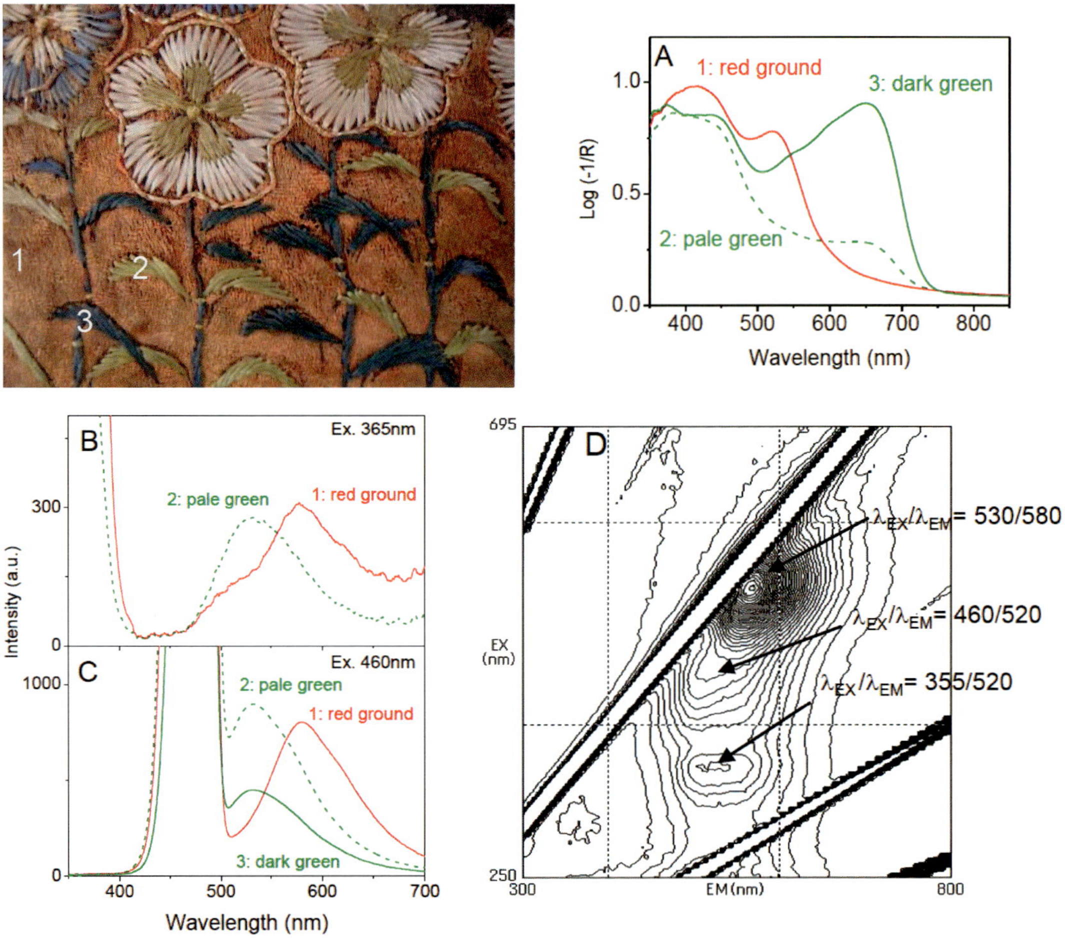

Figure 2 *Embroidered Memorial Altar Cloth with Flower Design*, early 17th century (KIT Collection AN. 416). Visible reflectance and fluorescence spectra of the red ground and pale and dark green embroideries. A: visible reflectance spectra; B: fluorescence spectra excited at 365 nm; C: fluorescence spectra excited at 465 nm; D: fluorescence EEM 3D spectrum of red ground.

around 650 nm without any fluorescence emission. The purple dye gromwell (*Lithospermum erythrorhizon* Siebold & Zucc.) was identified by its specific polynomial absorption, supported by its second derivative and absence of fluorescence.[11]

On-site analysis using a portable fibre-optic spectrometer

Satin Fragment with Design of Dragon, Cloud, and Wave on Blue Ground (KIT Collection A-c-8)

The first analysis example is *Satin Fragment with Design of Dragon, Cloud, and Wave on Blue Ground* (KIT Collection A-c-8) (Fig. 1), produced during the early-to-mid Qing dynasty in China. In this textile, spectroscopic analysis revealed that different dyes were used to represent the same colour. As shown in Figure 1, A and B, the red dyes in the dragon's ear, lip and nose (positions 1, 2 and 3) exhibited strong fluorescence emissions at 580 nm after excitation at 465 nm by the LED light source, which was typical for safflower, but no fluorescence emission was observed in the red dyes for the clouds beside the dragon (position 4). The second derivative of the visible reflection spectrum at position 4, without fluorescence, was almost identical with that of sappanwood, as shown in the inset of Figure 1A. Therefore the use of different dyestuffs (safflower and sappanwood) for seemingly the same colours could be determined easily using the portable device. Similar observations were made in the green *ungen*, which showed a stepwise fading of colour. The green *ungen* in the dragon's beard were achieved by double dyeing of the blue dye indigo, evidenced by typical absorption at 650 nm, with yellow dyes, as shown in Figure 1C. The yellow dye in the dark green showed no fluorescence, but the dyes of the other greens exhibited strong fluorescence emissions after excitation at both 365 nm and 465 nm (Fig. 1D), suggesting the presence of amur cork tree. Determination of non-fluorescent yellow dyes was difficult using only this portable device, with plausible options being pagoda tree and a pampas grass such as eulalia because they lacked the polynomial absorption spectra observed in gardenia.

Embroidered Memorial Altar Cloth with Flower Design (KIT Collection AN. 416)

The second example for on-site analysis is an *uchishiki*, *Embroidered Memorial Altar Cloth with Flower Design* (KIT Collection AN. 416) (Fig. 2). An *uchishiki* is an Edo era memorial altar cloth made by altering a *kosode* garment, associated with a deceased individual, donated to the temple. This textile has a typical design for the Keichō period of the early Edo era in Japan (1596–1615) with its red and black ground, and imprinted gold and fine embroideries. It is presumed to have been produced in the early 17th century.

The visible reflectance spectra and fluorescence spectra excited at 465 nm and 365 nm for three positions – the red ground (1), pale green (2) and dark green (3) embroideries – are shown in Figure 2, A–C. The reflectance spectra of the pale and dark green threads showed a characteristic absorbance at 650 nm corresponding to indigo, indicating that these green colours were made by multiple dyeing of a yellow dye with indigo. The yellow dyes of both green colours showed two absorbance peaks around 450 nm and 360 nm, with strong fluorescence after excitation at both 465 nm and 365 nm observed at 520–540 nm. The red ground also exhibited similar fluorescence after excitation with both wavelengths. These spectroscopic results were typical for dyestuffs including protoberberine, such as the amur cork tree. The red ground also exhibited another red dye with a 530 nm absorption and 590 nm emission, characteristic for safflower. This confirmed that both the red and green colours in the textile fragment were made by multiple dyeing using amur cork tree as the second dye component, combined with safflower for red and indigo for green. This composition was determined easily using a small spectroscopic device. To confirm these results, the fluorescence EEM 3D spectrum of the red ground was measured using a benchtop instrument fitted with an optical fibre. As shown in Figure 2D, three peak maxima were observed at $\lambda_{ex}/\lambda_{em}$ = 530 nm/580 nm, 355 nm/520 nm and 460 nm/520 nm, which were identical to those of multiple dyeing using safflower and amur cork tree. Thus, the results measured by the portable instrument were verified by the benchtop instrument.

Embroidered Fragments with Lotus Flower and Sanskrit Character from Banner for Buddhism Ceremony (KIT Collection AN. 486)

The third example is *Embroidered Fragments with Lotus Flower and Sanskrit Character from Banner for Buddhism Ceremony* (KIT Collection AN. 486) (Fig. 3). This fragment is from a set of 12 banners used in a temple ceremony and is thought to have been produced during the early Edo era in Japan (first half of the 17th century). Five colours – reddish brown, yellow, white, brown and blue – were observed in the embroidery of the lotus petals. The visible reflectance and fluorescence spectra excited at 465 nm and 365 nm for these coloured embroideries, obtained by a handheld portable device, are shown in Figure 3, and the spectral data are summarised in Table 2.

In both the yellow and white petals (positions 2 and 3), yellow dyes were detected with a strong fluorescence emission around 540 nm after excitation at 460 nm, characteristic for turmeric and/or amur cork tree. The reflectance spectra of the yellow petals showed absorption maxima at 360 nm and 460 nm, as well as a strong fluorescence emission excited at 365 nm, which indicated the presence of protoberberine alkaloids such as berberine, the major component of amur cork tree. The reflectance spectra of the yellow petal also showed another maximum absorption peak at 650 nm, indicative of the additional presence of indigo. It is probable that the current yellow colour was originally green from the multiple dyeing of yellow (amur cork tree) and blue (indigo). In comparison, the white petals showed only a single absorption at 450 nm and a

Table 2 Analytical results for *Embroidered Fragments with Lotus Flower and Sanskrit Character from Banner for Buddhism Ceremony* (KIT Collection An. 486).

Position	Present colour	Fluorescence		Reflectance	Dye identified	Proposed original colour
		Intensity λex(365)/ λex(465)[a]	Maximum emission (nm)	Maximum absorption (nm)		
1	reddish brown	–/–	–	530	sappanwood	red
2	yellow	++/+++	530	370, 450 650	amur cork tree indigo	green
3	white	+/+++	530	460	turmeric	yellow
4	brown	–/–	–	600	gromwell?	purple?
5	blue	–/–	–	650	indigo	blue

[a] Fluorescence intensity, excitation at 365 nm/465 nm: +++: very strong; ++: strong; +: moderate; ±: weak; –: not detected.

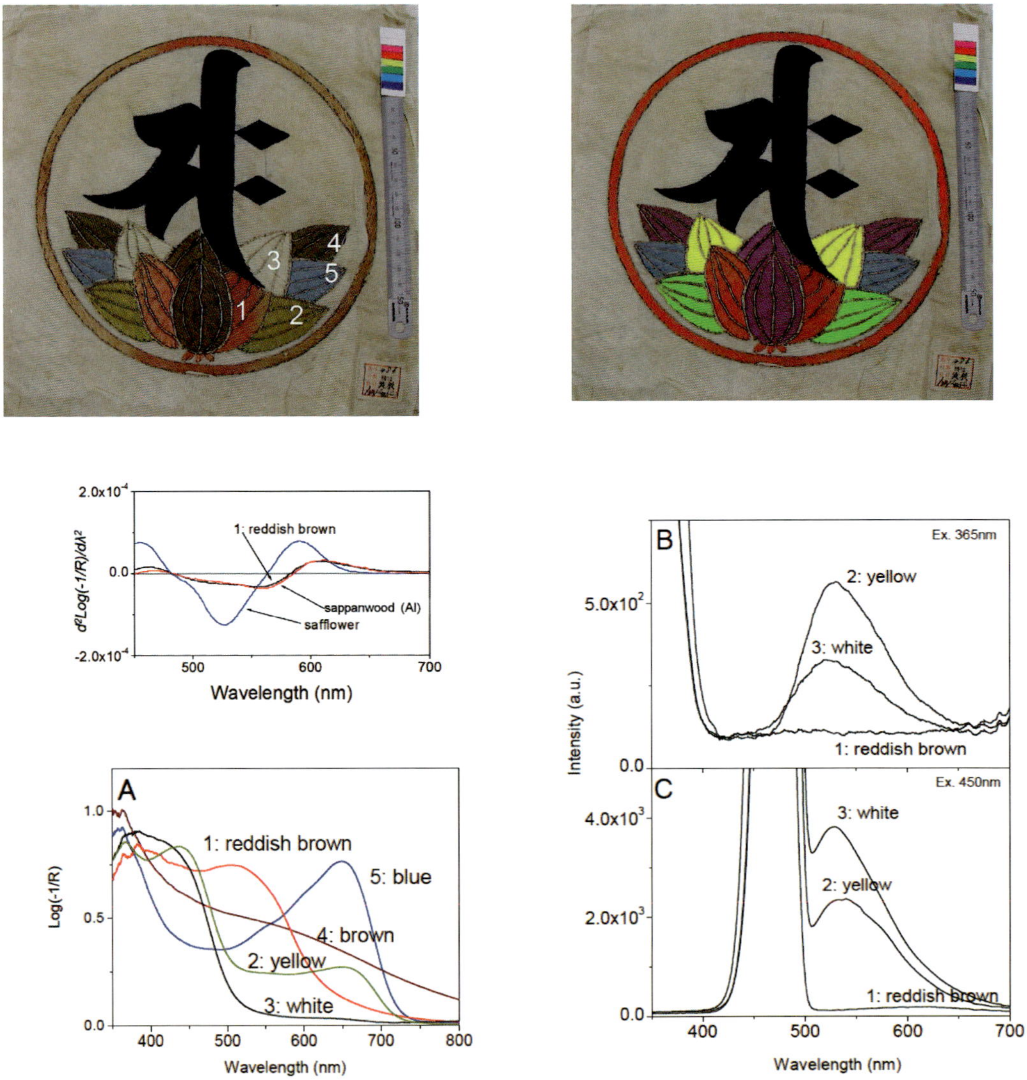

Figure 3 *Embroidered Fragments with Lotus Flower and Sanskrit Character from Banner for Buddhism Ceremony*, first half of the 17th century (KIT Collection AN. 486). Visible reflectance and fluorescence spectra. A: visible reflectance spectra and (above) their second derivatives for the red petal 1; B: fluorescence spectra excited at 365 nm; C: fluorescence spectra excited at 465 nm. Top left: original image; top right: reconstructed image based on the analysis.

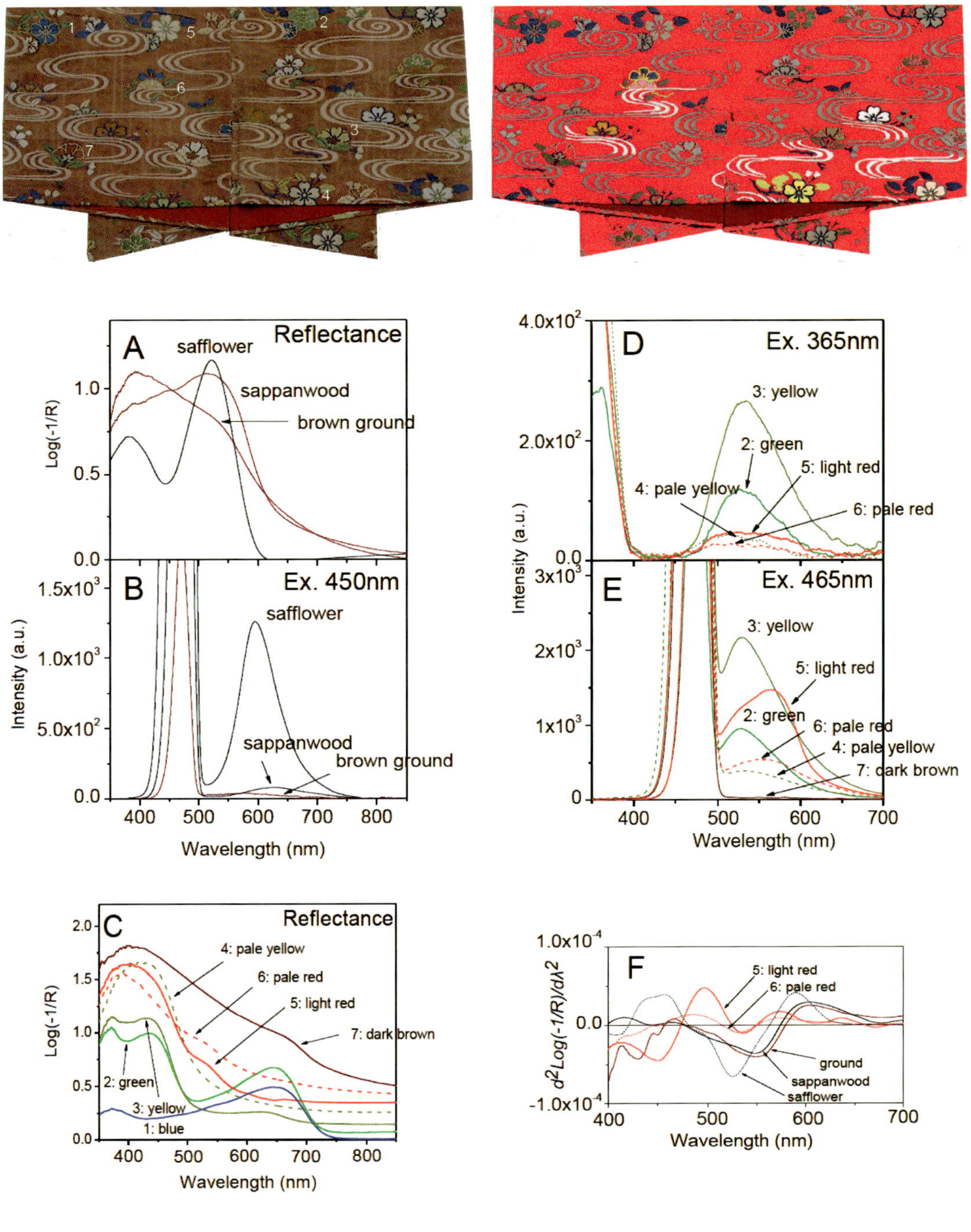

Figure 4 *Karaori with Design of Water Flow and Cherry Blossoms on Red Ground*, *c.*1700 (private collection). Visible reflectance and fluorescence spectra (excited at 465 nm and 365 nm) of the reddish-brown ground of the *karaori* costume. A: visible reflectance spectra of the red ground and red dye standards; B: fluorescence spectra of the red ground excited at 465 nm; C: visible reflectance spectra of the flower and leaf embroideries; D: fluorescence spectra of the flower and leaf embroideries excited at 365 nm; E: fluorescence spectra of the flower and leaf embroideries excited at 465 nm; F: second derivatives of visible reflectance spectra of red dyes for the ground and flowers (5 and 6). Top left: original image of whole costume; top right: reconstructed image of the lower side based on the analysis.

Table 3 Analytical results for *Karaori with Design of Water Flow and Cherry Blossoms on Red Ground* (private collection).

Position	Present colour	Fluorescence		Reflectance	Dye identified	Proposed original colour
		Intensity λex (360)/ λex (465)[a]	Maximum emission (nm)	Maximum absorption (nm)		
ground	light brown	–/±	(600)	536	sappanwood	red
1	blue	–/–	–	656	indigo	blue
2	green	++/+++	530	370, 438, 654	amur cork tree indigo	green
3	yellow	++/+++	530	370, 443, 645	amur cork tree indigo	pale green
4	pale yellow	–/±	(510)	430	eulalia	yellow
5	light red	–/+++	530(sh), 570	400, 530	turmeric safflower	orange
6	pale red	–/+	(570)	530	safflower	pink
7	dark brown	–/–	–	525	gromwell?	purple?

[a] Fluorescence intensity, excitation at 365 nm / 465 nm: +++: very strong; ++: strong; +: moderate; ±: weak; –: not detected.

weaker fluorescence emission excited at 365 nm than at 455 nm. These observations are characteristic for turmeric. The white colour of the petals therefore resulted from fading of the original yellow turmeric dyeing. The colour change from yellow to white occurred over a long period of ageing, probably due to light exposure. The reddish-brown petals (position 1) showed absorption at 530 nm but no fluorescence emission, characteristic for sappanwood and not safflower. The second derivative of the reflectance spectrum was also identical with that of typical sappanwood dyed with an alum mordant. In consideration of these analyses, the beautiful original image of 400 years ago was reconstructed as shown in Figure 3 (right-hand photograph).

Karaori with Design of Water Flow and Cherry Blossoms on Red Ground (private collection)

The fourth example for on-site analysis is a *karaori* Noh costume, *Karaori with Design of Water Flow and Cherry Blossoms on Red Ground* (Fig. 4), produced around 1700 in Japan (early Edo era). A *karaori* is a Japanese embroidery-like waving textile that originated in China as the *kara* and is used mainly as a Noh costume for women's roles. The present colour of the ground is a pale reddish brown and the brown linear back is the result of recent restoration.

The visible reflectance spectra of the reddish-brown ground showed an absorbance around 530 nm, but no fluorescence emission when excited at either 365 nm or 465 nm (Fig. 4, A and B). These observations were characteristic for sappanwood, a typical non-fluorescent red dye of ancient Japan; the second derivative of the reflectance spectrum was also consistent with this dye (Fig. 4F). Therefore the original colour of the ground was red, not the reddish brown seen in the present textile.

The flowers and leaves were strongly discoloured and appeared almost white, except for blue and green. The visible reflectance and fluorescence spectra for the threads of each design are indicated as positions 1–7 in Figure 4, C, D and E. Blue (1), green (2), and pale green (3) showed the typical absorption at 650 nm for indigo. Green (2) and pale green (3) also exhibited a strong fluorescence emission at 510 nm after multiple excitations at 365 nm and 465 nm, suggesting amur cork tree as a second dyestuff. Thus, the green and pale green were found to result from multiple dyeing of indigo and amur cork tree. The yellow threads (4) gave very weak fluorescence after excitation at both 365 nm and 465 nm, typical for the non-fluorescent yellow dye eulalia. Reddish yellow (5) and pale red (6) showed strong fluorescence at 590 nm, indicating safflower as the red dye. The reddish yellow (5) also showed fluorescence after excitation at 460 nm but not 350 nm; this is characteristic of the yellow dye turmeric. The damaged dark brown threads (7) showed an absorbance around 525 nm, but no fluorescence emission when excited at 465 nm (Fig. 4, C and E). These observations suggested that the damaged dark brown thread might be dyed with a purple dye, such as gromwell. The results are summarised in Table 3, and a reconstructed image simulated with typical colours for each identified dye is shown in Figure 4 (right-hand photograph).

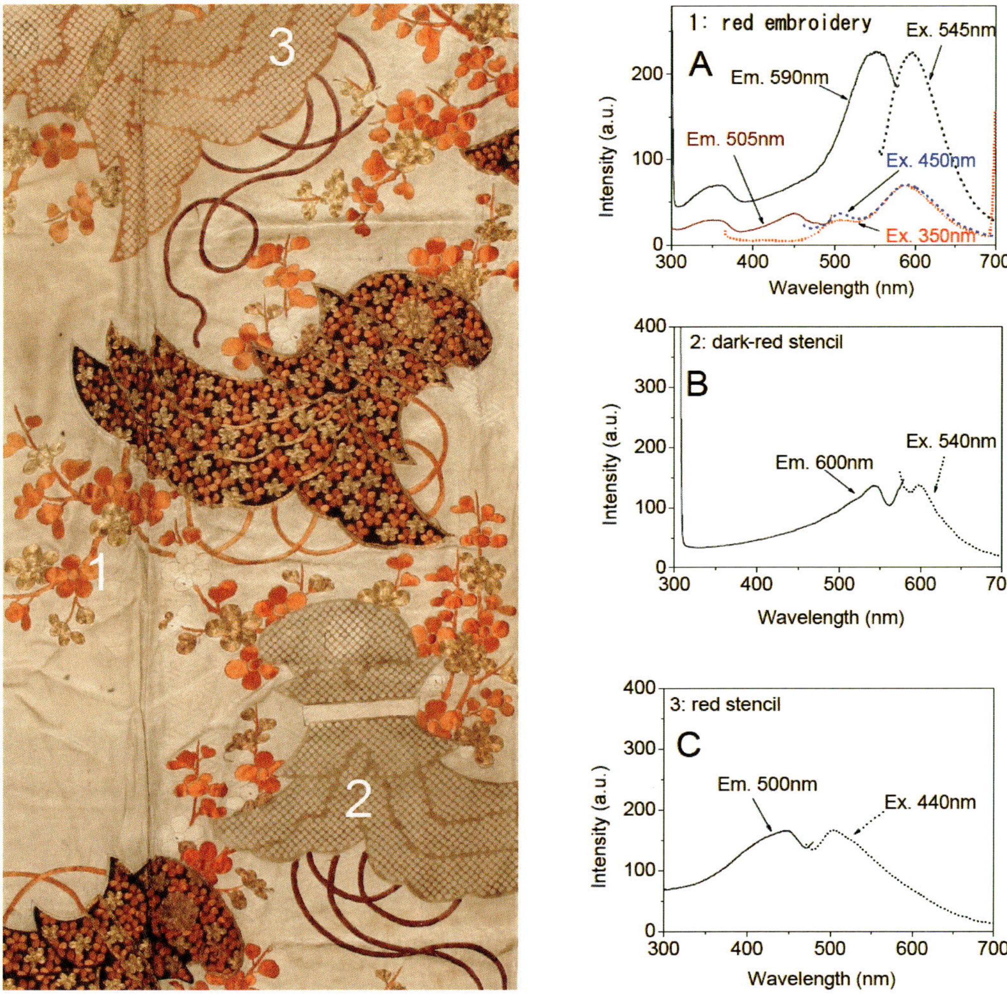

Figure 5 *Kimono Fragment with Design of Plum Flower, Bird, and Helmet on White Ground*, 18th century (KIT Collection AN. 395). Fluorescence spectra of embroideries and stencil prints. A: red embroidery, B: dark-red stencil print, C: scarlet stencil print.

Kimono Fragment with Design of Plum Flower, Bird, and Helmet on White Ground (KIT Collection AN. 395)

The last example is *Kimono Fragment with Design of Plum Flower, Bird, and Helmet on White Ground* (KIT Collection AN. 395) (Fig. 5) produced in Japan during the middle Edo era (18th century). The designs were created by red embroidery and two different red-coloured stencil prints (dark red and scarlet).

The fluorescence spectra of the red embroideries (Fig. 5A) and stencil prints (Fig. 5, B and C) are shown. The red embroideries, indicated as position 1, exhibited absorption spectra at 540 nm accompanied by strong fluorescence emission at 600 nm, providing evidence for safflower (Fig. 5A). Although the red stencil prints were distinguishable for two colours, dark red (position 2) and scarlet (position 3), both showed a peak at 540 nm, accompanied by no fluorescence emission around 600 nm. These spectroscopic characteristics are typical for sappanwood. In the scarlet stencil print, a yellow component in addition to the red colour was detected, and weak fluorescence indicated a yellow dye, excluding amur cork tree or turmeric, both of which show strong fluorescence. However, the dark red print did not contain any yellow component. The dark and scarlet colours of the stencil prints were easily distinguished by the presence of the yellow dye, which controlled the colour tone. Although determination of the yellow component was difficult using only the devices employed in the study, an Asian pampas grass as eulalia or the Japanese pagoda tree were considered as possibilities due to the polynomial characteristics of their absorption spectra.

Conclusion

This paper has described examples of non-destructive analysis using fibre-optic spectrometric devices for investigating the natural dyes and materials used for traditional 17th–19th-century East Asian (Japanese and Chinese) textiles. These devices yielded useful absorption and fluorescence spectral data, and information on multiple dyeing was easily obtained. These non-destructive methods were effective for determining the characteristics of traditional East Asian textiles, especially of precious textiles for which destructive methodologies are difficult to apply.

Editor's note and authors' addresses

*Author for correspondence: ysasaki@kit.ac.jp. At the time of writing the author was employed at Museum and Archives, Kyoto Institute of Technology, Japan. Paper received 13 March 2018; revised version received 16 September 2018. This paper was presented at *Dyes in History and Archaeology* 27, Istanbul, Turkey, 8–11 October 2008.

[1] Department of Fine Arts, Kyoto Saga University of Arts, 1 Saga-gotocho, Ukyo, Kyoto 616-8362, Japan
[2] Department of Biomolecular Engineering, Kyoto Institute of Technology, Matsugasaki, Sakyo-ku, Kyoto 606-8585, Japan

Acknowledgements

These works were partially supported by Grants-in-Aid for Scientific Research No. 20032010 and 19611003 from the Japanese Ministry of Education, Science, Culture, and Sports.

Notes and references

1. Sasaki, Y. and Sasaki, K. (2005) 'Visible spectroscopy, fluorescence spectroscopy, and mass spectrometry', in Nara National Research Institute for Cultural Properties (ed.), *Historic Silks, Culture, Construction and Conservation Science*, Tokyo, Kadokawa, pp. 156–68; Sasaki, Y. (2008) 'Cultural textiles under the scientific view point', *Journal of the Society of Fiber Science and Technology, Japan (Sen'i Gakka-ishi)* 64, pp. 36–44; Sasaki, Y., Sato, M., Koezuka, T., Kawai, T., Maekawa, Z. and Sasaki, K. (2001) 'Reflection spectrum analysis of natural dye for textile as cultural properties', *Journal of the Japanese Society for Scientific Studies on Cultural Property (Archaeology and Natural Science)* 40/41, pp. 1–15; Sasaki, Y., Sato, M., Koezuka, T., Nakajima, T., Maekawa, Z. and Sasaki, K. (2002) 'Mass spectroscopic identification of natural dyestuffs for ancient textiles', in *International Congress of the Conservation and Restoration for Archaeological Objects*, Nara National Institute of Cultural Properties, Nara, pp. 109–12; Sato, M. and Sasaki, Y. (2003) 'Blue and purple dyestuffs used for ancient textiles', *Dyes in History and Archaeology* 19, pp. 100–105; Sato, M. and Sasaki, Y. (2005) 'Studies on ancient Japanese silk fibres using FTIR microscopy', in R. Janaway and P. Wyeth (eds), *Scientific Analysis of Ancient and Historic Textiles: Informing Preservation, Display and Interpretation*, London, pp. 44–7; Sasaki, Y., Sato, M., Omura, M., and Sasaki, K. (2008) 'Dye analysis of braids used for Japanese armour dating from the 14th to the 16th centuries', *Dyes in History and Archaeology* 21, pp. 119–26; Sasaki, Y., Yano, T., Koike, T. and Sasaki, K. (2017a) 'Non-destructive dye analysis for the reconstruction of a 17th-century *Haori* fragment in the Tokugawa Art Museum', in J. Kirby (ed.), *The Diversity of Dyes in History and Archaeology*, London, pp. 257–64; Sasaki, Y., Matsubara, J. and Sasaki, K. (2017b) 'Red dye analysis for historical textiles with multiple spectroscopic methods: an example for Coptic textiles in the Nara Silk Road Exchange Museum', in J. Kirby (ed.), *The Diversity of Dyes in History and Archaeology*, London, pp. 107–15; Sasaki, Y. and Sasaki, K. (2017) 'Dye analysis of a 17th-century historic Japanese textile: a non-destructive approach', in J. Kirby (ed.), *The Diversity of Dyes in History and Archaeology*, London, pp. 265–77; Sasaki, Y. and Sasaki, K. (2009) 'Development of micro-analytical techniques for dyes in historic textiles (1) ESI mass spectrometric analysis of anthraquinone red dyes', *Journal of the Japanese Society for Scientific Studies on Cultural Property (Archaeology and Natural Science)* 59, pp. 83–95; Sasaki, Y., Kajitani, N. and Sasaki, K. (2009) 'Development of micro-analytical techniques for dyes in historic textiles (1) ESI mass spectrometric

analysis of the dyestuff used in 17th century Mughal carpets', *Journal of the Japanese Society for Scientific Studies on Cultural Property (Archaeology and Natural Science)*, 59, pp. 97–108; Sasaki, Y., Fukae, R. and Sasaki, K. (2010) 'Inorganic materials in ancient textiles (1): SEM-EDS analysis of inorganic pigments used for corrections of tie-dye dots', *Journal of the Japanese Society for Scientific Studies on Cultural Property (Archaeology and Natural Science)* 61, pp. 43–58; Sasaki, Y. and Sasaki, K. (2011) 'Development of micro-analytical techniques for dyes in historic textiles (3) ESI mass spectrometric analysis of early acid azo dyes', *Journal of the Japanese Society for Scientific Studies on Cultural Property (Archaeology and Natural Science)* 62, pp. 1–17; Sasaki, Y. and Sasaki, K. (2008) 'Dye analysis using portable spectroscopic systems for Japanese historical kimono', in 'Abstracts of 27th Meeting of Dyes in History and Archaeology, 8–11 October 2008, Istanbul, Turkey', p. 36.

2. Sasaki 2008 (see note 1 above).
3. Sasaki and Sasaki 2008 (see note 1 above).
4. Sasaki and Sasaki 2017 (see note 1 above); Matsuda, Y. and Miyoshi, M. (1989) 'Non-destructive analysis of ancient textiles', *Kobunkazai no Kagaku* [Journal of the Japan Society for the Conservation of Cultural Property] 34, pp. 1–10; Matsuda, Y. (1997) 'Non-destructive analysis of yellow natural dyes by three dimensional fluorescence spectroscopy', *Bunkazai Hozon Syufuku Gakkai-shi* [Journal of the Japan Society for the Conservation of Cultural Property] 41, pp. 54–63; Shimoyama, S. and Noda, Y. (1997) 'Non-destructive analysis of dyes in a Chinese brocade: determination of plant dyes in a 16th/17th-century textile by a three dimensional fluorescence spectrum technique with fibre optics', *Dyes in History and Archaeology* 15, pp. 70–84.
5. Sasaki *et al.* 2017a (see note 1 above); Sasaki *et al.* 2017b (see note 1 above); Sato and Sasaki 2003 (see note 1 above).
6. Sasaki and Sasaki 2017 (see note 1 above).
7. Sasaki and Sasaki 2017 (see note 1 above).
8. Goto, S. and Yamakawa, R. (1937) *Senryo Syokubutufu* [Japanese dye plants], Hakuo-sha, Kyoto, Japan; Uemura, R. (1986) *Engi-shiki kan* [Commentary for Rule of the Engi era (11th C) for the implementation of the penal code and administrative low], Tokyo, Iwanami.
9. Sasaki *et al.* 2009 (see note 1 above).
10. Sasaki and Sasaki 2017 (see note 1 above).
11. Sasaki *et al.* 2009 (see note 1 above).

Non-Destructive Spectroscopic Approaches for the Analysis of Dyes in Japanese Textiles: Analyses of an Early Edo Era (17th Century) Kimono Fragment

Yoshiko Sasaki[1*] and Ken Sasaki[2]

Abstract Detailed research on a 17th-century traditional textile fragment, *Fragment with Design of Tie-dye Dots* (hitta shibori) *and Embroidered Chrysanthemums, Pines, Bamboo, and Plums* (KIT Collection AN. 106), stored at the Kyoto Institute of Technology, Japan, was conducted by studying the dyeing and embroidery techniques and analysing the dyestuffs using non-destructive spectroscopic methods. The use of multiple dyestuffs was easily determined using fluorescence spectroscopy for the red ground with tie-dyed dots and the red embroidery thread. Yellow and purple embroidery threads were also characteristic for this textile. The yellow chrysanthemum embroideries on the left-hand side of the fragment were confirmed as having been dyed using amur cork tree, but the dyestuff used for the faded yellow colours on the right-hand side was turmeric. Similarly, the purple on the right-hand side was achieved using gromwell, while multiple dyeing of gromwell and indigo was used to achieve the colour on the left-hand side. By making numerous non-destructive spectral measurements across the textile fragment it was possible to demonstrate that similar colours in the embroidery were achieved using different techniques and materials, shedding light on the production of textiles in the early Edo era.

Introduction

Non-destructive approaches are important for chemical analyses of precious historical materials. In order to analyse natural and synthetic dyes used in these textiles, over recent years the authors have focused on the development of both non-destructive spectroscopic methods and mass spectrometric methods for the analysis of microsamples.[1] When employing non-destructive spectroscopic methods, a combination of analytical techniques is often required to increase the accuracy of identification. For example, dye analysis of Japanese kimono fragments (17th–18th century) using a combination of visible reflectance and fluorescence spectroscopy with infrared (IR) spectroscopy has been reported previously.[2] This paper assesses the suitability of fibre optic-based spectroscopic methods, including excitation-emission (EEM) matrix correlation fluorescence spectroscopy (generating three-dimensional fluorescence spectra),[3] for the analysis of historical textile dyes.

The fibre optic-based spectroscopic methods were applied to a typical 17th-century traditional Japanese textile fragment named *Fragment with Design of Tie-dye Dots* (hitta shibori) *and Embroidered Chrysanthemums, Pines, Bamboo, and Plums* (KIT Collection AN. 106), stored at the Kyoto Institute of Technology, Japan. This textile fragment was once part of a kimono, the rest of which is known as the *Koshimaki Fragment in Black Background Style with Chrysanthemums and Auspicious Motifs in Diagonal Cross Stripes* and is now part of a folding screen in the Nomura collection at the National Museum of Japanese History (H-35-16). The original kimono is thought to have

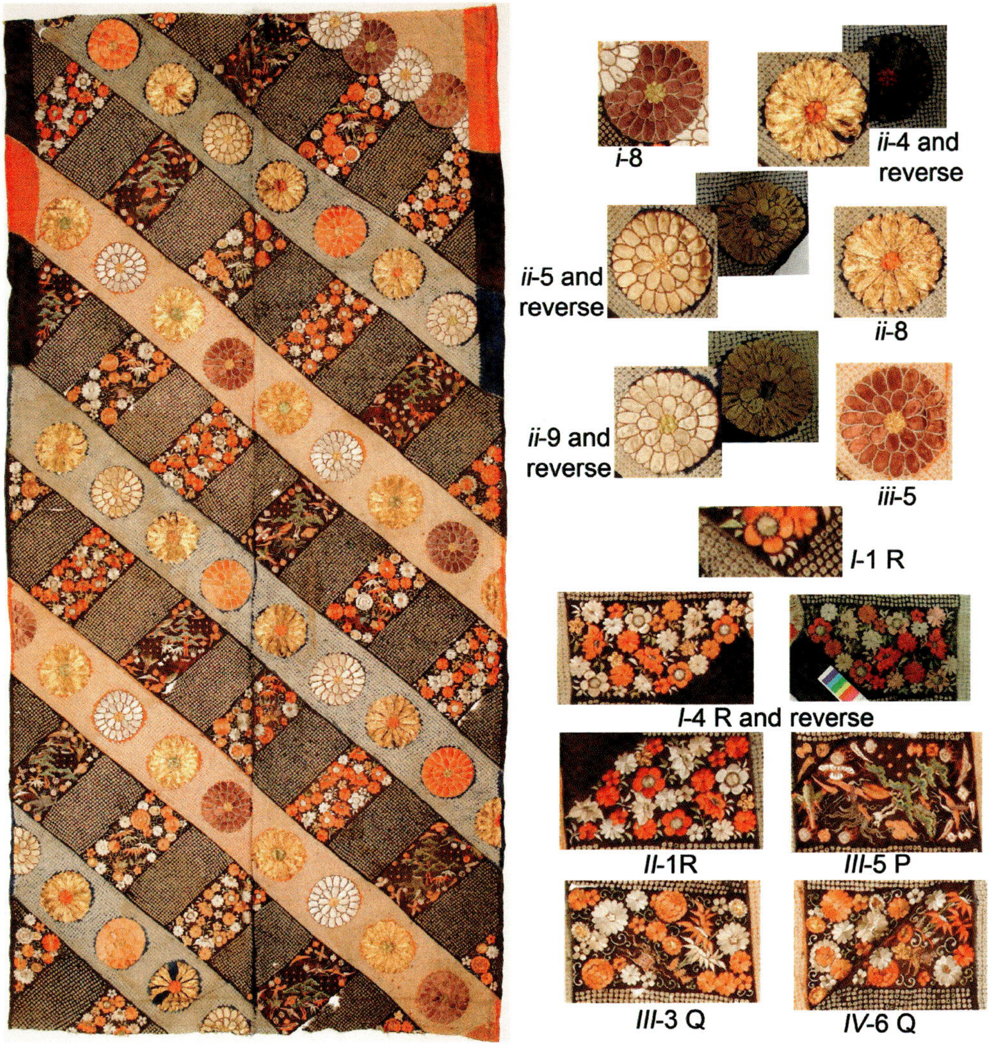

Figure 1 (*Left*) *Fragment with Design of Tie-dye Dots* (hitta shibori) *and Embroidered Chrysanthemums, Pines, Bamboo, and Plums* (KIT Collection AN.106). (*Right*) Details of the textile (see Figure 2 for numbering scheme).

belonged to Empress Tofukumon'in (1607–1678), granddaughter of Tokugawa Ieyasu, the first *shogun* of the Edo era.[4] Spectra were recorded at multiple locations across the textile fragment. In this paper, the detailed non-destructive dye analysis results for the whole textile fragment are presented and compared to those for the large chrysanthemum embroideries reported previously.[5] The analytical research on the dyestuffs was complemented by detailed study of the dyeing and embroidery techniques.

Experimental

EEM fluorescence spectra were recorded using a Hitachi F-4500 spectrofluorometer with a fibre-optic attachment. An Ocean Optics USB 4000 multichannel spectrometer with fibre optic was used as a portable device for the measurement of visible reflectance converted into absorbance spectra and fluorescence spectra. To clarify the characteristics for each dye, second derivatives of the visible reflectance spectra were calculated.

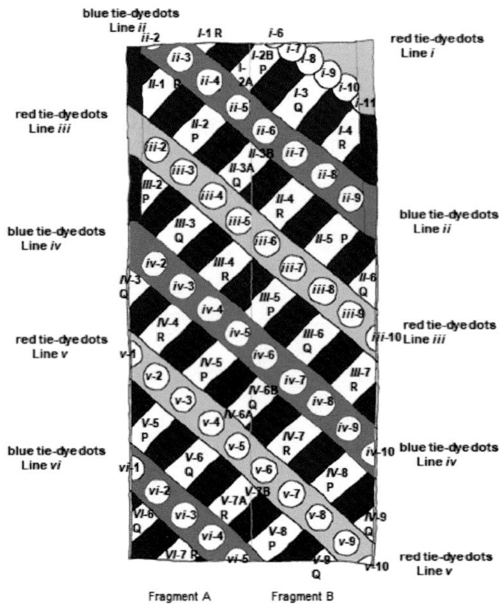

Figure 2. Arrangement and numbering of the embroideries of large chrysanthemums and auspicious and small flower motifs in the textile fragment (KIT Collection AN.106). R: small flowers; P: pine/crane/good luck treasures; Q: flower/bamboo/turtle motifs.

Results and discussions

Textile materials

Fragment with Design of Tie-dye Dots (hitta shibori) *and Embroidered Chrysanthemums, Pines, Bamboo, and Plums* (KIT Collection AN. 106) consists of two fragments sewn together (Fig. 1). The fragment measures 153 cm in length and 73–74 cm in width (73 cm at the top and 74 cm at the bottom). The textile consists of diagonal stripes decorated with large embroidered chrysanthemums against a ground of red or blue tie-dyed dots (with the background colours alternating). These stripes are separated by diagonal black stripes decorated with a repeating sequence of auspicious and small flower motifs separated by regions of black tie-dyed dots. The black stripes are arranged so that each of the auspicious and small flower motifs align and cross the stripes with embroidered chrysanthemums. The stripes of red and blue tied-dyed dots are 11 cm in width and are angled approximately 40 degrees from the horizontal. Eight full-size (7 cm in diameter) and two half-size embroidered chrysanthemums are arranged on each stripe.

The arrangement and numbering of the embroideries of the large chrysanthemums and auspicious and small flower motifs in the textile fragments are shown in Figure 2. The letters R, P and Q in the black ground of lines *I–VI* indicate the three auspicious and small flower motifs on the diagonal black stripes: small flowers, pine/crane/good-luck treasures and flowers/bamboo/turtles, respectively. Table 1a shows the distribution of the gold, red, yellow, white and purple colours of the embroideries of the large chrysanthemums on the stripes numbered *i–vi*. The observed techniques of embroidery and dyeing are characteristic of the late 17th century and consistent with textiles related to Empress Tofukumon'in.

Analysis of dyes used for the grounds with tie-dyed dots

The visible reflectance spectrum of the blue tie-dyed dots showed an absorption around 650 nm, characteristic for indigo (Fig. 3A). Dyestuffs used for the red tie-dyed dots were easily identified by their fluorescence spectra (excitation-emission matrix (EEM) and emission and excitation spectra). Strong fluorescence emissions of a major component at 580 nm (λ_{ex} = 540 nm) and of a minor component at 510 nm (λ_{ex} = 360 and 450 nm) were observed, as shown in Figure 3, B and D: the former emission was attributed to safflower and the latter to amur cork tree. These observations suggest that the yellow dye was used in the pre-dyeing process for red dyeing using safflower (Fig. 3C). No characteristic spectroscopic results were obtained for the black tie-dyed dots which were probably produced using carbon black.

Analysis of dyes used for the large chrysanthemum embroideries

The large chrysanthemum embroideries on the ground of the tie-dyed dots were also analysed using fluorescence spectroscopy. Examples of typical chrysanthemum embroideries are shown in Figure 1 (right). Safflower and amur cork tree

Table 1 Distribution of colour and dyes in the large chrysanthemum petals.

a. Colour of the large chrysanthemum embroideries

Line	Fragment A)					Fragment B)					
	1	2	3	4	5	6	7	8	9	10	11
i	-					P	W	P	W	P	W
ii	—	G	R	G	Y	G	R	G	W	—	
iii	—	G	W	G	P	G	W	G	P	G	—
iv	—	G	Y	G	R	G	W	G	R	G	
v	P	G	W	G	P	G	W	G	P		—
vi	Y	G	R	G	Y						

G: Gold thread, R: Red, Y: Yellow, W: White, P: Purple

b. Red large chrysanthemums

Line	Fragment A					Fragment B					
	1	2	3	4	5	6	7	8	9	10	11
i	-										
ii	—		S/A				S/A			—	
iii	—										—
iv	—				S/A				S/A		
v											—
vi			S/A								

S: Safflower, A: Amur cork tree

c. Yellow large chrysanthemums

Line	Fragment A					Fragment B					
	1	2	3	4	5	6	7	8	9	10	11
i	-										
ii	—				A				T	—	
iii	—										—
iv	—		A				T				
v											—
vi	A				A						

A: Amur cork tree, T: Turmeric

d. Purple large chrysanthemums

Line	Fragment A					Fragment B					
	1	2	3	4	5	6	7	8	9	10	11
i						G/I		G/I		G/I	
ii											
iii					G				G/I		
iv											
v	G				G				G/I		
vi											

G: Gromwell, I: Indigo

Yellow stitches of gold chrysanthemums

Line	Fragment A					Fragment B					
	1	2	3	4	5	6	7	8	9	10	11
i											
ii		A		A		T		T			
iii		A		A		T		A		T	
iv		A		A		T		T		T	
v		A		A		T		A			
vi		T		A							

A: Amur cork tree, T: Turmeric

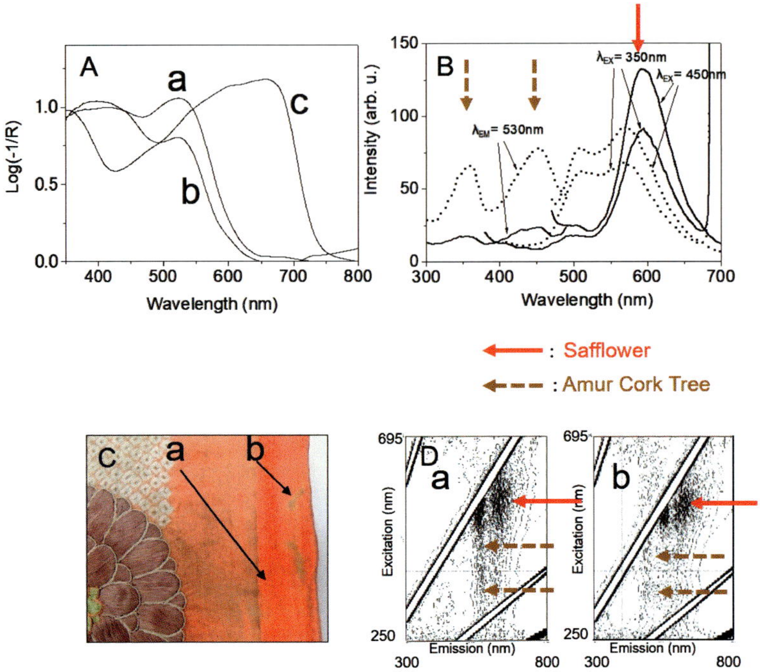

Figure 3 Visible reflectance and fluorescence (EEM and excitation and emission) spectra from the grounds with tie-dyed dots: a: red, b: red (discoloured), c: blue. A: Visible reflectance spectra; B: emission and excitation spectra; C: photo of red ground, top right of textile; D: excitation-emission matrix.

dyestuffs were identified in all the red embroidery threads examined based on the characteristic peaks ($\lambda_{ex}/\lambda_{em}$ = 535 nm/590 nm, 350 nm/510 nm and 460 nm/510 nm) in the EEM fluorescence spectra, as described above for the red tie-dyed dots. Spectra from the yellow chrysanthemum embroideries on fragment A (ii-5) exhibited only the two peaks characteristic of dyeing with amur cork tree (Fig. 4). By contrast, the dyestuff used for the faded yellow embroidery of fragment B (*ii*-9) was made with turmeric, as indicated by the characteristic peak in the EEM spectra ($\lambda_{ex}/\lambda_{em}$ = 460 nm/530 nm) (Fig. 4). Like the yellow embroideries, the yellow stitches of the gold embroideries were also dyed using amur cork tree or turmeric.

The purple embroideries of fragment A (*iii*-5) were dyed using gromwell, while both gromwell and indigo were identified in the purple colour of fragment B (*i*-8) using visible reflectance spectroscopy. The visible reflectance spectra and their second derivatives are shown in Figure 5, and the peak of indigo is indicated with an arrow. The distribution of the various dyes across all of the large chrysanthemum embroideries with red, yellow, and purple threads, as well as the yellow stitches of the gold, are shown in Table 1, b–d. As shown in these tables, similar colours (except for red) were produced using different dyestuffs in two fragments of the textile.

Analysis of dyes used for the auspicious and small flower motif embroideries

The auspicious and small flower motifs on the black diagonal stripes were made using eight colours: red, orange, pale pink and white for flower petals, pale yellow for flower centres, pale and dark green for pine and bamboo, and purple for cranes and turtles. These motifs were classified into three types: P, small flowers; Q, pine/crane/good-luck treasures; and R, flowers/bamboo/turtles. An example of each motif is shown in Figure 1 (right).

Analysis focused on a typical R motif (*I*-4). Typical red (*I*-4A), orange (*I*-4C), pale pink (*I*-4B) and white (*I*-4D) embroidery threads were measured to obtain visible reflectance and fluorescence spectra, as shown in Figure 6. Red, orange and pale pink showed absorptions around 530 nm and strong fluorescence emissions at 580 nm. The EEM spectra exhibited a peak at 535 nm/590 nm ($\lambda_{ex}/\lambda_{em}$), indicative of safflower. The red and orange threads also showed another emission peak at 510 nm when excited at 450 nm. The excitation spectra that yielded fluorescence emission at 510 nm indicated two absorption peaks at 350 nm and 460 nm, a result confirmed by the EEM spectra, which indicated two peaks at 350 nm/510 nm and 460 nm/510 nm ($\lambda_{ex}/\lambda_{em}$). These results were attributed to the presence of amur cork tree as a second dye component in these colours. Because there was evidence for the fading of the colours of the threads from red to orange, both the red and orange threads examined are believed to have originally been the same colour (red) but one has discoloured to orange after ageing.

Fluorescence spectra for the pale pink threads indicated no yellow components, indicating that they were dyed with safflower alone. The white threads did not display any characteristic spectra, suggesting they were undyed. Both the pale and dark green threads, as exemplified by motif P (*III*-5) on fragment B, showed a characteristic absorption in the visible reflectance spectra at 650 nm attributed to indigo, as well as the specific fluorescence emission at 510 nm (when excited by light at both 355 nm and 460 nm) of amur cork tree. Thus, these green threads were found to have been produced by dyeing using both indigo and amur cork tree. Two types of fluorescence spectra were observed in the pale yellow flower centres as shown in Figure 7. Emission at 530 nm after excitation at 460 nm was observed in the centre of *I*-4 on fragment B, characteristic for turmeric. However, the pale yellow flower centre of *I*-1 on fragment A exhibited the characteristic excitation spectra of amur cork tree.

The purple cranes and turtles were determined to have been dyed using gromwell on the basis of their visible reflectance spectra and the second derivatives of these spectra. On fragment A, as exemplified by the turtles in *III*-3, gromwell was used alone, but on fragment B (*IV*-6B), an additional absorption maximum at around 650 nm indicated the use of indigo (Fig. 8, the peak of indigo is indicated with an arrow). Therefore, as

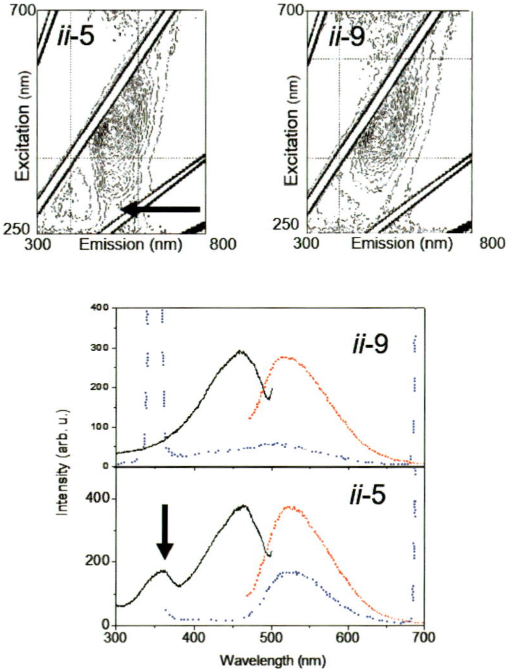

Figure 4 Fluorescence spectra (EEM and excitation and emission) spectra of yellow large chrysanthemum embroideries, *ii*-5 and *ii*-9. Excitation spectra monitored at 520 nm. Red lines: emission spectra excited at 450 nm. Blue lines: emission spectra excited at 350 nm.

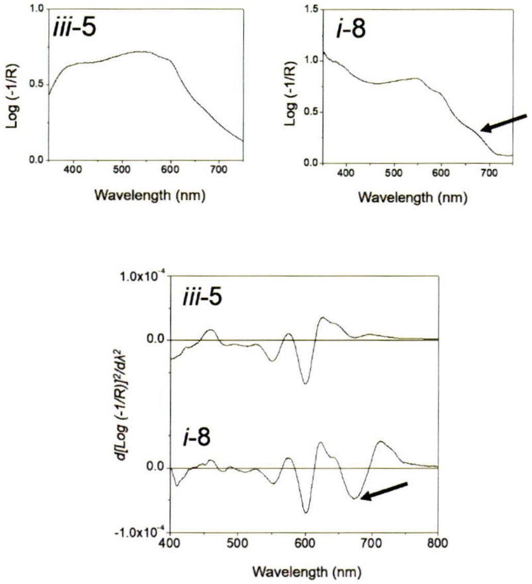

Figure 5 Visible reflectance spectra and their second derivatives for the purple large chrysanthemum embroideries, *i*-8 and *iii*-5.

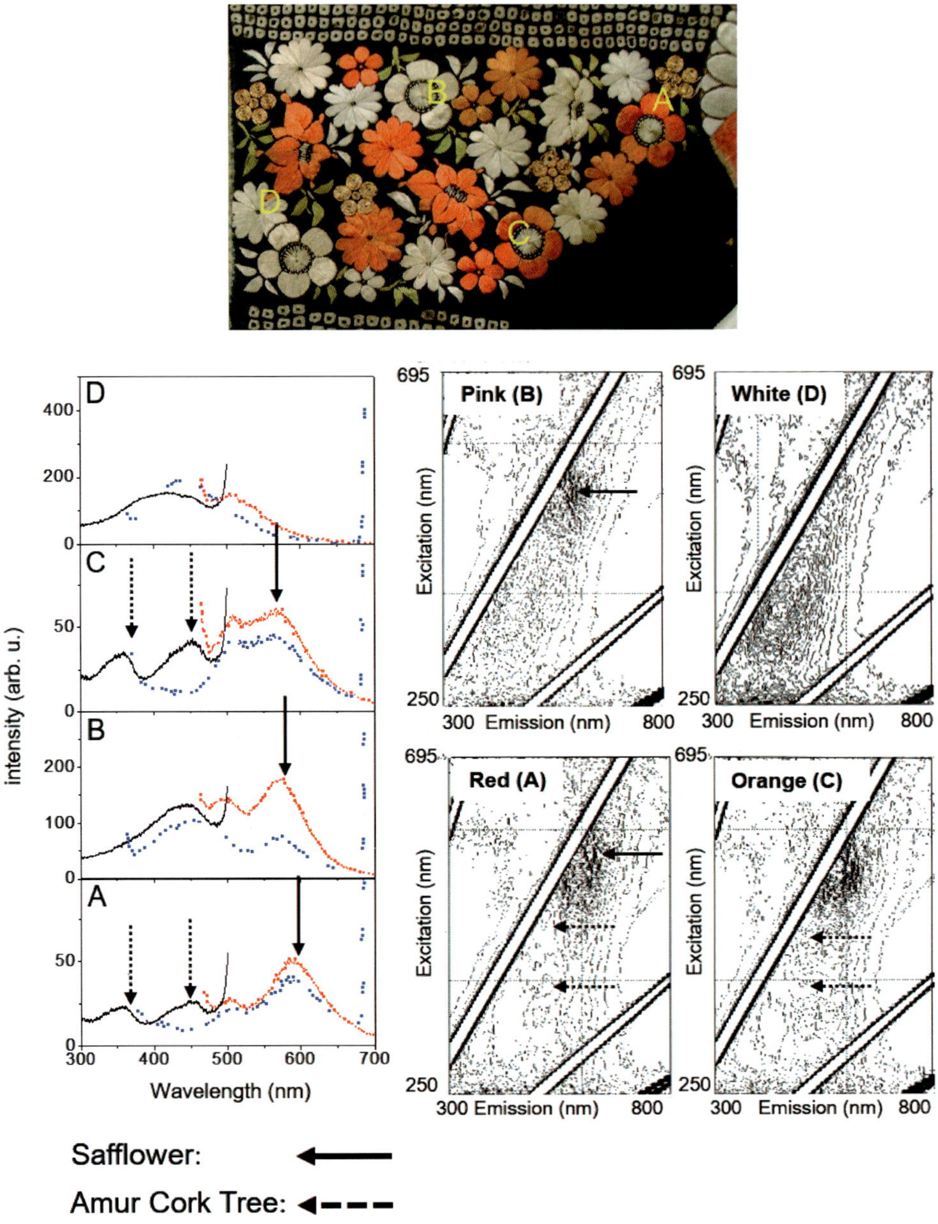

Figure 6 (*Top*) Measurement locations for motif *I*-4. EEM (*lower right*) and excitation-emission (*lower left*) fluorescence spectra of small flowers from motif *I*-4. A: red petal (*I*-4A); B: pale pink (*I*-4B); C: orange petal (*I*-4C); and D: white petal (*I*-4D). (*Lower left*) Solid lines: excitation spectra monitored at 520 nm. Red lines: emission spectra excited at 450 nm. Blue lines: emission spectra excited at 350 nm.

found in the large chrysanthemum petals, different dye distributions were used for similar purple-coloured threads in the auspicious and small flower motifs in fragments A and B.

All the yellow centres of the small flowers and the purple colours in the cranes and turtles were analysed (see Table 2). The distribution of the two different yellow dyes, amur cork tree and turmeric,

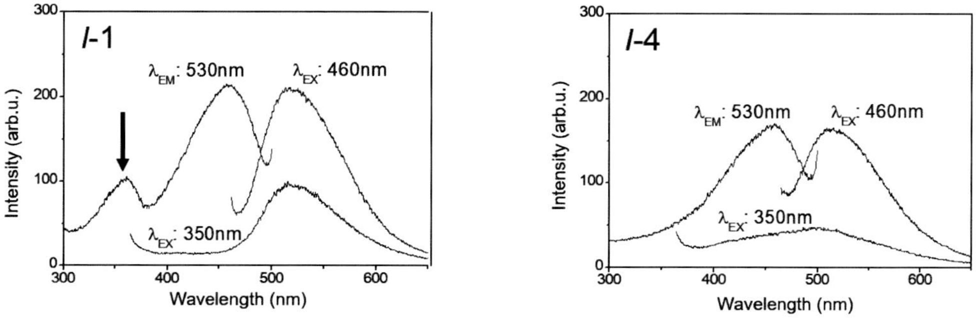

Figure 7 Fluorescence excitation and emission spectra of small flower centres (yellow), *I*-1 and *I*-4.

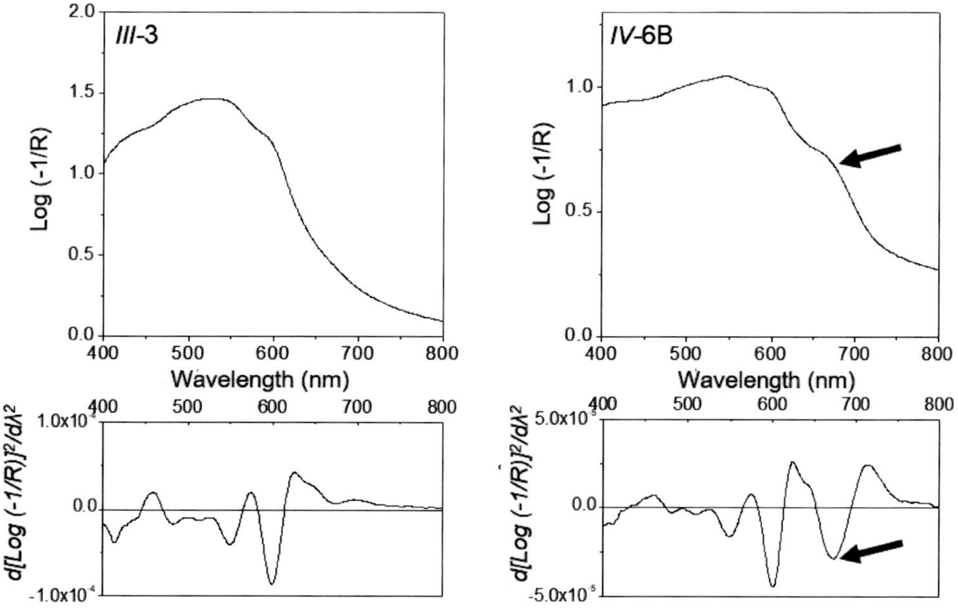

Figure 8 Visible reflectance spectra and their second derivatives for the purple cranes and turtles (*III*-3 and *IV*-6B).

across the two fragments differs. All the yellow centres of the small flowers in the right-hand fragment B were dyed using turmeric, whereas in fragment A, the yellow dyes consisted of a mixture of turmeric and amur cork tree. Similarly, the purple threads also indicated two types of dye: gromwell only (fragment A) and double dyeing of gromwell and indigo (fragment B). Therefore, as in the large chrysanthemum embroideries, different dye distributions for fragments A and B were observed in the small flower centres, with different dyes being used to produce similar colours. The colours of the small flowers will, of course, have changed due to ageing over several hundred years, but based on these dye analyses, a reconstruction of the original colours may be proposed (Fig. 9).

Technical insights into dyeing and textile production

The red colours used in the tie-dyed ground, the large chrysanthemums and small flowers were produced by multiple dyeing using safflower and amur cork tree. The yellow or orange colours in the ground or small flowers are likely to result from the discoloration of

Table 2 Dye distribution in the yellow centres of the small flowers and the purple cranes and turtles.

Motif	P	Q	R	P	Q	R	P	Q	R
	Fragment A		Fragment B						
Line *I*	*I*-1 A	*I*-2A – *I*-2B –	*I*-3 –	*I*-4 T	–				
Line *II*	*II*-1 T	*II*-2 G	*II*-3A – *II*-3B –	*II*-4 T	*II*-5 –	*II*-6 –	–		
Line *III*	–	*III*-2 G	*III*-3 G	*III*-4 T	*III*-5 G/I	*III*-6 –	*III*-7 T	–	
Line *IV*	–		*IV*-3 –	*IV*-4 –	*IV*-5 G	*IV*-6A – *IV*-6B G/I	*IV*-7 T	*IV*-8 –	*IV*-9 –
Line *V*	–				*V*-5 G	*V*-6 –	*V*-7A A *V*-7B T	*V*-8 G/I	*V*-9 –
Line *VI*	–					*VI*-6 –	*VI*-7 A	–	
	Fragment A							Fragment B	

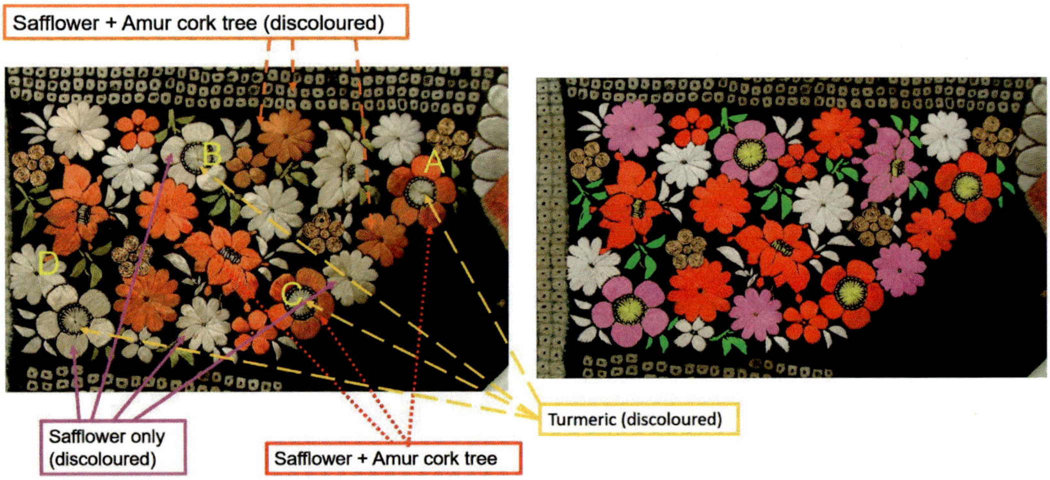

Figure 9 (*Left*) Summary of dyestuffs identified in the small flower motif (*I*-4). (*Right*) Reconstruction showing the likely original appearance of the small flower motif (*I*-4).

the safflower component. Multiple dyeing was also observed in a previous study of a *kosode* fragment of the *Kanbun* period (1675–92),[6] and it is believed that this would have been the general procedure used for producing a scarlet colour in the early Edo era. By contrast, no yellow component was observed in the pale pink embroidery. Therefore, the red colour tone would appear to have been controlled by the addition of a yellow dye component.

In fragments A and B, different dyeing methods were used for similar yellow and purple colours. In the purple colours of the large chrysanthemums and

Table 3 Current and probable original colours in the large chrysanthemum decorations based on dyestuff analysis.

Present colour	Green	Red	Orange	Yellow	White	White	White	Purple
Dyes identified	Indigo/ amur cork tree	Safflower/ amur cork tree	Discoloured safflower/ amur cork tree	Amur cork tree	Discoloured safflower	Discoloured turmeric	None	Gromwell or gromwell/ indigo
Proposed original colour	Green	Red	Red	Yellow	Pink	Yellow	White	Purple

Table 4 Embroidery techniques for the flower centres for the large chrysanthemum decorations.

a. Flower centre stitch

Line	**Fragment A**					**Fragment B**				
	1	2	3	4	5	6	7	8	9	10
i	–	–	–	–	–	–	cording	non-split float	cording	non-split float
ii	–	–	cording	split satin	cording	non-split float	cording	non-split satin	non-split float	–
iii	–	split satin	cording	split satin	cording	non-split float	cording	split satin	cording	–
iv	–	split satin	cording	non-split satin	cording	non-split float	cording	non-split float	non-split satin	non-split satin
v	cording	split satin	cording	split satin	cording	non-split satin	cording	non-split float	non-split satin	non-split satin
vi	cording	non-split float	cording	non-split satin	cording	–	–	–	–	–

b. Flower centre rim stitch

Line	**Fragment A**					**Fragment B**				
	1	2	3	4	5	6	7	8	9	10
i	–	–	–	–	–	–	–	outline	–	outline
ii	–	–	–	outline	–	outline	–	fixing	fixing	
iii	–	outline	–	outline	–	fixing	–	outline	–	–
iv	–	outline	–	outline	–	outline	–	outline	fixing	fixing
v	–	outline	–	outline	–	fixing	–	fixing	fixing	fixing
vi	–	outline	–	outline	–	–	–	–	–	–

turtle/crane designs, simple dyeing by gromwell was used for fragment A whereas multiple dyeing with gromwell and indigo was employed for fragment B. There were also characteristic differences between the fragments in terms of the yellow colours. Turmeric was used mainly for the yellow dye of the gold threads in the large chrysanthemums and the centre of the small flowers in fragment B (Table 1, b–d and Table 2). By contrast, amur cork tree was used for the yellow dye of all the yellow large chrysanthemums and most of the yellow stitches of the gold threads in fragment A. Both amur cork tree and turmeric were combined for the small flower centres in fragment A. The yellow large chrysanthemums of fragment B now appear almost white due to fading of the turmeric dyestuff and it is likely that the original colour would have been a similar yellow to that produced by amur cork tree in fragment A. The proposed original colours of the dyes based on the analytical results are summarised in Table 3. If account is taken of the fading of colours, the original systematic colour arrangement of the large chrysanthemums is revealed.[7]

The variation in the dyes used in the two fragments is also mirrored by observed differences between fragments A and B in terms of the stitch techniques (Table 4). The two fragments are believed to have been produced individually by two different groups employing differing embroidery techniques and dyed threads. Thus, spectroscopic analyses of this textile fragment provided important and novel knowledge on the production of Japanese textiles in the early Edo era, complementing other technical evidence.

Conclusion

Spectroscopic approaches combining fluorescence and reflectance spectra with their second derivatives have been shown to be suitable for the identification of dyes from textiles. By employing benchtop instruments with fibre optics, these analytical methods become entirely non-destructive and permit multi-point analysis to be undertaken, even on rare cultural textiles. These methods were applied successfully to a precious textile fragment from the 17th century (KIT Collection AN.106) that was examined thoroughly by making spectroscopic measurements at more than 200 points; this would have been difficult to conduct using destructive (microsampling) analysis. Multi-point analysis also made it possible to determine that different dyes (or dye combination) were used for similar colours, lending further support to the suggestion that the two parts of the textile fragment examined were produced by two different groups. The particular dye combinations identified also allow conclusions to be drawn concerning early Edo period dyeing technology and textile production.

Editor's note and authors' addresses

*Author for correspondence: ysasaki@kit.ac.jp. At the time of writing the author was employed at Museum and Archives, Kyoto Institute of Technology, Japan. Paper received by 13 March 2018; revised version received 16 September 2018. This paper was presented at *Dyes in History and Archaeology* 28, Poznań, Poland, 21–24 October 2009.

[1] Department of Fine Arts, Kyoto Saga University of Arts, 1 Saga-gotocho, Ukyo, Kyoto 616-8362, Japan

[2] Department of Biomolecular Engineering, Kyoto Institute of Technology, Matsugasaki, Sakyo-ku, Kyoto 606-8585, Japan

Acknowledgements

This research was partially supported by Grants-in-Aid for Scientific Research No. 20032010 and 19611003 from the Japanese Ministry of Education, Science, Culture, and Sports.

Notes and references

1. Sasaki, Y., Matsubara, J. and Sasaki. K. (2017) 'Red dye analysis for historical textiles with multiple spectroscopic methods: an example for Coptic textiles in the Nara Silk Road Exchange Museum', in J. Kirby (ed.), *The Diversity of Dyes in History and Archaeology*, London, pp. 107–15; Sasaki, Y. and Sasaki, K. (2017) 'Dye analysis of a 17th-century historic Japanese textile: a non-destructive approach', in J. Kirby (ed.), *The Diversity of Dyes in History and Archaeology*, London, pp. 265–77; Sasaki, Y., Fujii, K. and Sasaki, K. (2010) 'Research on the textile fragment (AN. 106) stored in Kyoto Institute of Technology, "Fragment with Design of Tie-dye Dots (*hitta shibori*) and Embroidered Chrysanthemums, Pines, Bamboo, Plums, and Good-luck Treasures"', *Bunkazai Hozon Syufuku Gakkai-shi* [Journal of the Japan Society for the Conservation of Cultural Property] 55, 38–66; Sasaki, Y. and Sasaki, K. (2011) 'Development of micro-analytical techniques for dyes in historic textiles (3) ESI mass spectrometric analysis of early acid azo dyes', *Journal of the Japanese Society for Scientific Studies on Cultural Property (Archaeology and Natural Science)* 62, pp. 1–17; Sasaki, Y. and Sasaki, K., 'A non-destructive approach to dye analysis for typical 17th–19th-century East Asian textiles', in this volume, pp. 39–49.
2. Sasaki *et al.* 2017 (see note 1 above).
3. Sasaki *et al.* 2017 (see note 1 above); Matsuda, Y. and Miyoshi, M. (1989) 'Non-destructive analysis of ancient textiles', *Kobunkazai no Kagaku* [Journal of the Conservation of Cultural Property] 34, pp.1–10; Matsuda, Y. (1997) 'Non-destructive analysis of yellow natural dyestuffs by three dimensional fluorescence spectroscopy', *Bunkazai Hozon Syufuku Gakkai-shi* [Journal of the Japan Society for the Conservation of Cultural Property] 41, pp.54–63; Shimoyama, S. and Noda, Y. (1997) 'Non-destructive analysis of dyes in a Chinese brocade: determination of plant dyes in a

16th/17th-century textile by a three-dimensional fluorescence spectrum technique with fibre optics', *Dyes in History and Archaeology* 15, pp. 70–84.
4. Collection number H-35-16 in the National Museum of Japanese History, Sakura, Japan; Nomura, S. (1938) *Jidai Kosode Hinagata Byobu* [Folding screen of ancient Kosode], Kyoto, Geirin-do, 1, p. 16.
5. Goto, S. and Yamakawa, R. (1937) *Senryo Syokubutufu* [Japanese Dye Plants], Kyoto, Hakuo-sha; Uemura, R., 1986, *Engi-shiki kan* [Commentary for Rule of the Engi era (11th century) for the implementation of the penal code and administrative law], Tokyo, Iwanami.
6. Nomura 1938, (see note 4 above).
7. Sasaki, Y., Fukae, R. and Sasaki, K. (2010) 'Inorganic materials in ancient textiles (1): SEM-EDS analysis of inorganic pigments used for corrections of tie-dye dots', *Journal of the Japanese Society for Scientific Studies on Cultural Property (Archaeology and Natural Science)* 61, pp. 43–58.

The Tradition of Using Natural Dyes in Latvia

Anete Karlsone

Abstract This paper provides an introduction to the development of the tradition of using dye plants in Latvia after the discovery of aniline dyes and offers an insight into the written sources from various periods on their use. The public started taking an interest in the use of dye plants at the close of the 19th century. The economic hardships that followed both the First and Second World War stimulated interest in dyeing with natural dyes and encouraged the practice of this craft. During the interwar period in independent Latvia, amateur ethnographers, artists and home economics teachers revived this disappearing aspect of traditional culture. This renaissance has continued and is still evolving.

Introduction

Recent years have seen a growing interest in the use of plant dyes among handicraft workers in Latvia as evidenced by various public activities: for example, many master classes in dyeing are held each year and natural dyes have been the subject of several exhibitions. The travelling exhibition *Nature Dyes* was displayed at the Latvian Museum of Natural History in Riga from January to March 2012, attracting huge numbers of visitors. During the next four years, the exhibition was shown at different venues throughout Latvia, introducing visitors to the dyes available in nature and their use in dyeing textile fibres, especially woollen thread. Another touring exhibition was held at the regional centres of Limbaži in September–October 2017, presenting objects made by the local weaving and knitting community using woollen thread dyed with plant dyes, and in February 2018, it travelled to another regional centre, Alūksne. Dyeing thread with plant dyes has been included as an activity at several public events.

Public interest has also stimulated research in this field, although such studies are still limited.[1] The tradition of dye plant use in Latvia has received very little archaeological or ethnographic study: an extensive body of material, collected by earlier generations of researchers, has only been partially studied to date.[2] This paper aims to provide an introduction to the development of the practice of dyeing with plants in Latvia following the discovery of aniline dyes, and to offer an insight into the written sources from various periods that have influenced the current use of these plants. This article utilises ethnographic material in various document repositories – such as, for example, the Department of Ethnography, National History Museum of Latvia and the Repository of Ethnographic Material, Institute of Latvian History, University of Latvia – and in the Latvian press from the 1880s onwards and academic publications. Comparative material includes Latvian folklore (e.g. folk songs) as well as publications on archaeological research in Latvia.

The earliest evidence of the use of dyes in present-day Latvia

The earliest evidence of the use of dye plants in the region corresponding to present-day Latvia comes from textile fragments recovered from archaeological sites. The oldest of these date from the 3rd–4th century AD.[3] Chemical analyses of textile colorants discovered in the course of archaeological excavation have only been carried out for blue colours, revealing the presence of the dye indigo.[4] Red, yellow and brown colours have also been detected in the ancient textiles but unfortunately no analysis of these colorants has been conducted to identify the plant sources of these dyestuffs. The origin of the brown colour remains unclear: it may have been obtained by using pigmented brown wool or by dyeing of white wool. Alternatively the woollen thread may have become brown as a result of prolonged burial. A multidisciplinary group of scientists at the University of Latvia plans to continue research on this question.

The Latvian public started taking an interest in the use of dye plants at the close of the 19th century when Latvia was still part of the Russian Empire. As in the other major empires of Europe, processes of nation building and self-awareness were under way. In this context, attention focused on exploring and gathering together the cultural heritage of the people. In addition to the recording of folklore – for example, folk songs, beliefs, melodies – the newly emerged community of Latvian intellectuals also took an interest in traditional skills, including the dyeing of textile fibres using natural dyes.

A publication dating from 1885 was the first to record ethnographic information on dyeing with plant dyes.[5] The author, Pēteris Blaus (pseudonym *Zilais*, which means 'Blue'), was a Latvian writer, teacher and folklore collector who compiled his manuscript on this topic as early as 1876. He recorded descriptions of dyeing by elderly women, mentioning a total of six dye plants and including brief accounts of the dyeing processes. The information provided in this account could possibly date as far back as the beginning of the 19th century or even the end of the 18th century.

Additional ethnographic descriptions of dyeing with plant dyes were published around the turn of the 20th century (1893, 1894, 1902),[6] recorded by individuals with a special interest in this subject, possibly collectors of Latvian folklore. They also listed the plants used for dyeing, indicated the colours obtained and gave brief accounts of the dyeing processes. These publications mention a total of 19 plants including native plants such as dyer's chamomile (*Anthemis tinctoria* L.), silver birch (*Betula pendula* Roth) and alders (*Alnus* spp. Mill.) as well as imported dyestuffs including logwood (*Haematoxylum campechianum* L.), brazilwood (*Paubrasilia echinata* (Lam.) E.Gagnon, H.C.Lima & G.P.Lewis, syn. *Caesalpinia echinata* Lam.), indigo obtained from *Indigofera* spp. L., turmeric (*Curcuma longa* L.) and walnut (*Juglans* spp.). These scientific names were assigned by later scholars.

The publications on dyeing with plant dyes by Miķelis Skruzītis

In the ethnographic material published in the late 19th and early 20th century, all the dye plants and other substances are referred to by their popular, not scientific, names. Later authors using this material in their publications have, in many cases, added their own interpretations of the initial record, supplementing it with the Latin names of the plants and additional information. The first author to cite the Latin names of the dye plants in his publications was the Latvian ethnographer Miķelis Skruzītis, who also possessed a knowledge of chemistry. Papers by Skruzītis on dyeing with plant dyes were published in Latvian in 1895, 1902 and 1916, and in German in 1900.[7] In his articles, the information on dye plants and their use was presented in an ordered system for the first time; the dyeing methods were collected together and the plants used were grouped scientifically.

Skruzītis not only made use of previously published information – he also collected additional evidence including interviewing people on the subject of dyeing with plant dyes. In order to determine the precise names of the plants in Latvian, German and Latin, Skruzītis consulted German lexicons and German-Latvian dictionaries as well as recent botanical discoveries made by Latvian botanists.[8] He also paid particular attention to the description of dyeing methods, seeking to generalise them and to explain in chemical terms, for example, fermentation with groats or other substances, and the use of lye obtained from ash. It is significant that, in contrast to many other authors who have written about dye plants, Skruzītis may also have had practical experience of the dyeing process.

The articles by Skruzītis mention a total of 26 dye plants, in most cases indicating the colour that could be obtained and the method of dyeing. His publications clearly describe the status of the use of plant dyes in late 19th-century Latvia, when local plants were no longer used to obtain blue. At this time woad (*Isatis tinctoria* L.), previously cultivated in the region of present-day Latvia as indicated by folkloric evidence, had already been forgotten. Blue was now obtained by using indigo sourced from *Indigofera* species and imported from India.

Ethnographic evidence collected by the Board of Monuments

After 1918, when the independent state of Latvia was established, researchers continued to study the use of natural dyes. Information was collected during the ethnographic expeditions organised by the Board of Monuments in the years 1924–1931, and in 1942 and 1943. The records of this fieldwork are currently stored at the Ethnography Department of the National History Museum of Latvia. They preserve an extensive body of evidence regarding the natural dyes and dyeing methods used in the late 19th century and the first decades of the 20th century. These documents indicate that in most of Latvia some individuals retained knowledge of how to use dye plants and that this knowledge continued to be applied in practice.[9]

The descriptions compiled in the course of ethnographic expeditions not only provide information on dyes and their uses, but also indicate the extent of the use of this traditional skill during the time period as recalled by the interviewees. In analysing the material collected by the Board of Monuments, it is important to consider the manner in which these people discussed plant dyes. Three degrees of precision may be distinguished:

1. The speaker uses verbs in the present tense: 'you collect ...', 'you pour over...' etc. Such a mode of expression indicates a person who is not only very familiar with the process being described but also that they have practical experience of it, perhaps even comparatively recently.
2. The speaker uses verbs in the past tense: 'were gathered ...', 'was dyed ...' etc. This suggests someone who is describing past practices, either observed or carried out at an earlier time, perhaps in their youth. This raises the possibility that the account may be imprecise or even erroneous due to memory loss.
3. The speaker uses verbs in the past tense as reported speech: 'they are said to have dyed ....'. This suggests that the described activities may have almost disappeared and are no longer being practised. They might not have been actually witnessed by the speaker, who may have obtained the information from accounts by other people.

Actual samples of thread or dried plants included in several of the records from the Board of Monuments expeditions indicate a high degree of credibility. These samples prove that plant dyes were still being used and demonstrate that those who provided the information were actively using the process. More frequent, however, are records of activities carried out at some non-specific, earlier time; these may have simply been observed without attention being paid to significant details by the person providing the information.

In the records of the Board of Monuments ethnographic expeditions, the use of dye plants is mentioned more than 300 times with specific colours indicated in 260 cases. The palette is very broad – green, brown and yellow colours are more commonly cited but the full colour spectrum is represented. Importantly, the colours red, blue and vivid black were obtained predominantly from imported natural dyes: brazilwood (*Paubrasilia echinata* (Lam.) E.Gagnon, H.C.Lima & G.P.Lewis), indigo extracted from *Indigofera* L. species and possibly also logwood extract from *Haematoxylum campechianum* L.

Scientific names can be assigned to 42 of the dye plants listed in the records, which also mention colorants of mineral and animal origin. Sixteen plants appear frequently (recorded at least five times), some of which are not cited in earlier sources, including:

- The northern Labrador tea (*Rhododendron tomentosum* Harmaja, syn. *Ledum palustre* L.): the above-ground parts of the plant were used for beige, brown and grey.
- The common oak (*Quercus robur* L.): the bark and sometimes leaves were used for beige, brown, greenish brown, reddish brown and black.
- The broad bean (*Vicia faba* L.): the above-ground parts were used for black, bluish black, brown, grey and dark green.

Twenty-seven plants were recorded as dye plants for the first time, whereas six of the plants known from earlier ethnographic expedition documents are not mentioned at all. Surprisingly, the plants not mentioned at this time include clubmosses (*Lycopodium* spp.) even though they occupy an established place among dye plants according to older as well as more recent sources.

In addition to dye plants, the records of the Board of Monuments ethnographic expeditions also include descriptions of various dyeing techniques. However, not all of these techniques can be regarded as accurate – in some cases the inconsistencies, which could have occurred for various reasons, are quite easy to recognise. In other cases, however, an error in the record can only be revealed when dyeing experiments are conducted.

Publications on dyeing with plant dyes in the interwar period

Most of the interwar publications on dyeing with plant dyes that appeared in the Latvian press were authored by handicraft workers, home economics teachers and instructors of courses given by the children's agricultural education organisation Mazpulks (the Latvian equivalent of '4-H', an organisation based in the USA dedicated to developing children's potential). Amateur ethnographers also contributed articles on this topic. This information appeared in women's magazines, in publications on ethnography and cultural history and in regional newspapers. Many of these publications simply give a list of dye plants, supplemented with notes on dyeing techniques. Significant for the current research project are the dyeing recipes sent in by people from various parts of Latvia, in which they describe procedures that have actually been used for dyeing with natural dyes. These recipes may thus be regarded as providing ethnographic data which supplement the main body of the ethnographic knowledge of dye plant use. The information disseminated by teachers and home economics instructors drew not only on earlier published material but also on the authors' own experiences of conducting experiments on dyeing.

In 1931, the magazine *Zeltene* also published an extensive general discussion of dyeing with plant dyes, written by ethnographer Jānis Niedre.[10] Detailed study of the information published by Niedre reveals that in some instances he had rather freely interpreted or 'improved' the descriptions in the Board of Monuments records. This has been established only very recently through careful comparison of the published material with the original documents.

The relatively large number of lists of dye plants and recipes for their use in the interwar publications evidences the increased public demand in Latvia for information of this kind. Even though the number of dye plant species listed in any one publication may be few, the total number of botanically identifiable taxa had increased to 92. The plants listed most frequently in the publications of this period include:

- The alders, black alder (*Alnus glutinosa* (L.) Gaertn.) and grey alder (*A. incana* L.) Moench): bark and leaves for brown, black, beige, reddish-brown and greenish colours.
- Silver birch (*Betula pendula* Roth): leaves for yellow and green, cambium for reddish hues and buds for brown.
- Common heather (*Calluna vulgaris* (L.) Hull): the above-ground parts of the plant for green, brown-green, yellow, grey-green and brown.
- Dyer's chamomile (*Anthemis tinctoria* L.): flowers for yellow and green.
- Various bedstraw species (*Galium* spp. L.): roots for red and brownish red.
- Alder buckthorn (*Frangula alnus* Mill.): bark, fruits and leaves for yellow, reddish, brown and green.

- Common oak (*Quercus robur* L.): bark, leaves and acorns for grey, black, brown, reddish hues, beige and bluish grey.

Overall 42 plants, comprising almost half of the total, are mentioned for the first time in the publications from the interwar period. If these were authored by a home economics teacher, the plants may have been added to the list as a result of experimentation, suggesting that the tradition of using dye plants was continuing to develop and the range of plants used was being extended through practical testing.

The publications of the 1920s and 30s also provide an indication of the motivation behind the practice of using dye plants at a time when chemical dyes predominated in textile dyeing. The following considerations are mentioned:

- Artistic/aesthetic: pleasant, mild colours; the great variety of hues; correspondence of the colour range to the local landscape; every colour in harmony with the others.
- Practical: the durability of the dye on an item in use – it does not run or fade; the low cost of the dye.
- Cultural/historical: an ancient, inherited skill (starting from 1934 it is emphasised that this *is* an ancient Latvian skill).
- Ecological: dye plants as an untapped natural resource.

In the interwar period, instruction in the use of dye plants was included in school curricula and in the public activities of both pupils and women's organisations. These endeavours testify to efforts to re-establish the tradition.

Martha Bielenstein: 'The Ancient Latvian Dyeing Methods'

The Latvian dyeing tradition was a subject of interest not only to Latvians but also to Baltic Germans. In 1935, *Die altlettischen Färbmethoden* (The Ancient Latvian Dyeing Methods) by Martha Bielenstein was published in Riga in German.[11] The author carefully assembled and systematised the information available at the time, which was quite chaotically dispersed in various sources. Like her father, the linguist, cultural historian and Lutheran pastor August Bielenstein, Martha made extensive use of linguistic evidence and Latvian folklore in her discussion of the historic use of plant dyes. In addition to utilising her father's research findings,[12] she was strongly influenced by the article 'On the dyes used for ancient Latvian dress' by folklorist and ethnographer Pēteris Šmits,[13] and by the publications of the abovementioned Miķelis Skruzītis in which folk song and other sources were employed to characterise the use of dyes.[14] In some cases, however, she interpreted the poetic folklore texts rather too literally or in a romantic light.

In accordance with the research approaches of the time, Martha Bielenstein focused on identifying the layers that were likely to be the oldest in the material under study. Adhering to the migration theory current in the late 19th and early 20th century, she sought to ascertain the locales where the use of dye plants and other cultural phenomena originated. Aiming to reconstruct the earlier history of textile production in present-day Latvia, she gathered as much information as possible on this topic. For example, she collected evidence on the utilisation of linden bast fibre in clothing and in the production of fabrics for everyday use such as bed linen. Her findings may represent the sole source of information on this very archaic stage in the history of Latvian textiles.

In her research, Martha Bielenstein consulted studies by linguists and archaeologists from various countries, as well as major dictionaries and works by botanists in Latvia and abroad. She used one of the first lists of medicinal plants in Latvian, compiled by the pharmacist Ernests Birzmanis.[15] She also relied on the publications of the first Latvian botanist, Jānis Ilsters.[16] For her general work on dye plants and dyeing methods, she used almost all the published ethnographic documents available at the time, as well as unpublished material, citing a total of about 90 publications. For various reasons, an accurate list of the literature she consulted has not yet been compiled, but most of these publications are known. They comprise a very large range of material, especially considering that Bielenstein had no university education.

In addition to her extensive study of the academic literature, Bielenstein included in her book information she obtained from some 20 individuals living in Latvia and abroad. She listed 112 dye plants as well as mineral pigments in her book, supplying not only the popular names in Latvian and German but also the scientific names (following the method of classification employed at that time). This facilitates identification of most of the plants listed. She also provided a systematic overview of the mordants used in earlier times and the techniques for their use, distinguished local from imported mordants, and described in considerable detail the various traditional methods of dyeing.

A translation of Martha Bielenstein's book into Latvian and English is currently being prepared for publication that will also include the original German text with a scientific commentary, supplemented with research articles on Bielenstein and the traditional uses of dye plants in Latvia.[17]

Ethnographic material on dye plants in Latvia after the Second World War

From the outbreak of the Second World War, particularly in 1942–43, the number of publications on dye plants appearing in the Latvian press increased as a result of the economic conditions which prompted people to turn to traditional skills, including those required for using dye plants, in order to maintain a decent standard of living. After the Second World War, under Soviet rule, the collection of ethnographic material continued. In the context of research expeditions, information on dyeing with natural dyes was also recorded.[18] This documentation mainly addresses the traditional methods of using dye plants as practised in Latvia in the 20th century. It supplements ethnographic material in commercial publications from the first half of the 20th century and provides new information on this topic for the second half of the century.

Overall, the evidence regarding dye plants obtained during this ethnographic fieldwork spans the period from 1947 to 1984. This type of information had seldom been recorded previously as dyeing with plant dyes was not a research focus in its own right. When data were collected on other topics – for example, dress production, weaving, knitting and other women's handicrafts – incidental information pertaining to dyeing with plant dyes was also included. Because dyeing for home use was generally a female activity, the respondents were usually women. The evidence for dyeing with plant dyes collected during this fieldwork applies to the period from about the 1880s – the oldest respondents, interviewed in 1947–48, were born in the latter half of the 1860s – up to the mid-1980s. The respondents often recalled activities of their mother or grandmother during their childhood. Especially prominent are records from both the First and Second World War, when dyeing with plant dyes was practised more intensively.

Compared to the number of dye plants recorded in the documents covering the preceding periods, the number mentioned in the ethnographic material held at the Institute of Latvian History is considerably reduced – only 39 plants are named. Unquestionably, the most frequently used plant is alder (bark and leaves for grey, brown, black, beige to reddish brown and green), but the species used – black alder (*A. glutinosa* (L.) Gaertn.) or grey alder (*A. incana* (L.) Moench) – is rarely indicated. Next in frequency of use are onion skins (*Allium cepa* L.) for beige, yellow to brown, reddish and green colours followed by common heather (*Calluna vulgaris* (L.) Hull), used for yellow, green and brown and birch (*Betula* sp.). It should be noted that the dye sources from birch include the outer bark and cambium for grey, beige and brown hues, which is mentioned relatively infrequently in earlier sources, as well as the leaves for yellow and green.

The next most frequently employed plants are juniper (*Juniperus communis* L.) – the bark, and the branches with the needles for green – and sometimes the fruit of alder buckthorn which were used for yellow and green. The production of dark indigo blue by fermentation of the dye in a urine vat is also recorded. The available records do not indicate whether the purchased colorant in such cases was a natural or synthetic indigo dye. However, all the descriptions refer to the ancient, traditionally used technique of fermentation in a urine vat. Last among the more frequently used dye plants are the cones of spruce (*Picea abies* (L.) H.Karst.) for brown, and the bark of the common oak (*Q. robur* L.) for brown, grey and green colours.

In all these documents, the hues created most frequently by dyeing with natural dyes are browns, greens, greys and yellows. The plants most commonly utilised were those that allowed the desired result to be achieved using a straightforward dyeing technique. Complex dyeing techniques and plants that were more difficult to obtain, including those that gave brighter colours, generally fell out of use in the second half of the 20th century; they appear relatively infrequently in the records. Vivid reds and blues, which are more difficult to achieve using natural dyes, were more often obtained from synthetic dyes, which gave the desired result more easily and quickly. When plant dyes were used for everyday needs, the colours that were easily obtained were sufficient because they suited the styles in demand for dress and interior textiles of the period. Additional artistic qualities were provided by creating nuanced intermediate hues, which would have been more difficult to achieve using the synthetic dyes.

Interest in the use of plant dyes increased in Latvia in the late 1970s and continued up to the early 1990s due to the influence of handicrafts teacher and ethnographer Ilga Madre. She collected ethnographic material on dyeing with plant dyes, conducted numerous experiments and promoted this ancient skill through various practical activities as well as television appearances. She also authored a book on plant dyes, which may still be found on the bookshelves of many handicraft workers and is widely used.[19]

Conclusion

Information in written sources on the use of dye plants in the region of present-day Latvia can be found starting from the end of the 19th century, and by the turn of the 20th century, a total of 134 plant sources for natural dyes had been recorded. The economic hardships of both the First and Second World War created a demand for ancient skills which had not been entirely forgotten, engendering a keen interest in the practice of dyeing with local plant dyes and prolonging its application. Knowledge of this skill was thus passed on to the next generation. In independent Latvia during the interwar period, researchers sustained an interest in the use of plant dyes and this aspect of traditional culture was revitalised by amateur ethnographers, artists and home economics teachers. In the 20th century, the number of dye plants employed increased through practical experimentation; however those plants that permitted the required colour to be obtained by relatively simple means were used more frequently. The specific lists of dye plants used during different periods varies somewhat but the tradition of dyeing with plant dyes continues, and is still evolving, in Latvia today.

Editor's note and author's address

Address for correspondence: Institute of Latvian History, University of Latvia, Kalpaka bulvāris 4, Riga LV-1050, Latvia; anete.karlsone@gmail.com.
Paper received by 13 March 2018; revised version received by 16 September 2018. This paper was presented at *Dyes in History and Archaeology* 36, Hampton Court, UK, 26–28 October, 2017.

Acknowledgements

The preparation of this article was funded as part of the University of Latvia research project no. ZD2015/AZ85, 'The territory of Latvia as a zone of contact between different cultural spaces, religions, and political, social and economic interests from prehistory up to the present day'.

Notes and references

1. Karlsone, A. (2017) 'Baltvāciete Marta Bīlenšteina un viņas pētījums par latviešu tradicionālo kultūras mantojumu' [The Baltic German Martha Bielenstein and her research into the Latvian traditional culture heritage], in I. Saleniece *et al.* (eds) *Vēsture: Avoti un cilvēki: Humanitārās fakultātes XXVI starptautisko zinātnisko lasījumu materiāli. Vēsture XX.* [Proceedings of the 26th International Scientific Readings of the Faculty of Humanities. History XX] vol. 20, Daugavpils, Saule Academic Publishing House, pp. 167–73; Karlsone, A. (2016a) 'Krāsaugu lietojuma tradīcijas atspoguļojums latviešu preses izdevumos (19. gadsimta beigas – 1940. gadi)' [Reflection of the tradition of using plant dyes in Latvian press (end of the 19th century to the 1940s)], in I. Saleniece *et al.* (eds) *Vēsture: avoti un cilvēki. Daugavpils Universitātes Humanitārās fakultātes 25. starptautisko zinātnisko lasījumu materiāli. Vēsture XIX.* [Proceedings of the 25th International Scientific Readings of the Faculty of Humanities. History XIX] vol. 19, Daugavpils, Saule, pp. 125–31.; Karlsone, A. (2016b) 'Rakstītie

avoti par krāsaugu lietojuma tradīciju Latvijas teritorijā. 19. gs. beigas-1944. gads' [The written sources on the tradition of use of plant dyestuffs in the territory of Latvia in the period between the end of the XIX century and the year 1944], in A. Vilcāne (comp.) *Arheoloģija un Etnogrāfija,* vol. 29, Riga, Zinātne, pp. 193–204.
2. Auškāps, J. (1940) 'Dažu tautas lietāto krāsojumu izturība' [The durability of some folk-used dyes] *Latvijas Universitātes Raksti. Ķīmijas fakultātes sērija* 6, pp. 101–13; Ozols, K. (1939) 'Senie latviešu audumi' [Ancient Latvian fabrics], *Senatne un Māksla* 2, pp. 76–101; Niedre, J. (1931) 'Seno latviešu krāsas un krāsošana stādvielām' [Ancient Latvian colours and dyeing with plant dyestuffs], *Zeltene* nr. 21–24; Skruzītis, M. (1895) 'Latviešu tautas apģērbs savā vēsturiskā attīstībā un nozīmē. Kultūrvēsturiska studija no M. Skruziša' [Latvian folk clothing in its historical development and meaning. Cultural historical studies from M. Skruziša], *Austrums* 1, pp. 10–13; Skruzītis, M. (1902) 'Par latviešu senējo krāsošanas mākslu' [About Latvian old dyeing art], *Tēvija* nr. 21–23; Skruzītis, M. (1916) 'Par latviešu cittreizējo krāsošanas mākslu' [About Latvian dyeing skills in other times], *Līdums* nr. 202–205; Bielenstein, M. (1935) *Die altlettischen Färbmethoden: Studien zur idg. Altertumskunde.* [The Ancient Latvian Dyeing Methods: Studies in Indo-European Antiquity] Riga, Ernst Plates.
3. Žeiere, I. (2014) 'Latviešu senais apģērbs' [Latvian ancient clothing], in J. Stradiņš (ed.) *Latvija un latvieši* 3, Riga, Latvijas Zinātņu akadēmija, vol. 1, p. 377.
4. Ozols 1939 (see note 2 above), p. 76.
5. Zilais *pseud.* (i.e., Blaus, P.) (1885) 'Par dzīparu madarošanu un sieviešu apģērbu agrākos laikos' [About yarn dyeing and women's clothing in earlier times], *Rīgas Latviešu biedrības Zinību komisijas rakstu krājums*, vol. 3, Riga, pp. 77–80.
6. Saržants, Kr. (1893) 'Par krāsošanu' [About dyeing], *Etnogrāfiskas ziņas par latviešiem. Dienas Lapas pielikums* 4, pp. 58–9; Zilumkalns, A. (1894) 'Par krāsām un krāsošanu' [About colours and dyeing], *Etnogrāfiskas ziņa par latviešiem. Dienas Lapas pielikums* 6, p. 93; Lindulis *pseud.* (i.e. Linde, J.) (1902) 'Mūsu senču pervēšanas māksla' [Our ancestral colouring art], *Talsu kalendārs 1903. gadam*. pp. 40–42.
7. Skruzītis 1895, 1902, 1916 (see note 2 above); Skrusits, M. (1900) 'Über die ehemalige lettische Färbekunst' [About the earlier Latvian art of dyeing], *Sitzungsbericht der Altertumsgesellschaft Prussia* 21, Königsberg, Im Selbstverlage der Gesellschaft, pp. 199–216.
8. Ilsters, J. (1883) *Botanika tautas skolām un pašmācībai* [Botany for Folk Schools and Self-study] Riga, Pūzīšu Ģederts; Ilsters, J. (1884) 'Latviešu botāniski nosaukumi, krāti un sastādīti no J. Ilstera' [Latvian Botanical Names, Chapters compiled by J. Ilster], *Rīgas Latviešu biedrības Zinību komisijas rakstu krājums*, vol.2, pp. 63–81; Ilsters, J. (1885) 'Latviešu botāniski nosaukumi, sastādīti no J. Ilstera. Otrais salasījums' Ilstera' [Latvian botanical names drawn from J. Ilster. Second Edition], *Rīgas Latviešu biedrības Zinību komisijas rakstu krājums*, vol. 3, pp. 68–74; *Meyers Konversations-Lexikon* (1839–1855), vol. 46+6, Hildburghausen; Ulmann, C.Ch. (1872) *Lettisches Wörterbuch. Erster Theil. Lettisch=deutsches Wörterbuch.* Riga.; Ulmann, C.Ch. and Brasche, G. (1880) *Lettisches Wörterbuch. Zweiter Theil. Deutsch=lettisches Wörterbuch*, Riga, Leipzig.
9. LNVM ZAE – Scientific Archives of the Ethnographic Department of the History Department of the Latvian National Museum of History; ZAE 47 – Document set: Dyeing, washing and bleaching.
10. Niedre 1931 (see note 2 above).
11. Bielenstein 1935 (see note 2 above). Reprinted as Bielenstein, M. (1994) *Kleidung und Farbmethoden der alten Letten*, Michelstadt, Neuthorverlag.
12. Bielenstein, A. (1907) *Die Holzbauten und Holzgeräte der Letten: ein Beitrag zur Ethnographie, Kulturgeschichte und Archaeologie der Völker Russlands im Westgebiet*, St. Petersburg, Kaiserlichen Akademie der Wissenschaften.
13. Šmits, P. (1923) 'Par seno latviešu apģērba krāsu' [The colour of ancient Latvian clothing], *Filologu biedrības raksti*, vol. 3, Riga, pp. 7–33.
14. Skruzītis 1895, 1092, 1916 (see note 2 above).
15. Birzmanis, E. (1897) *Latvijas ārstniecības stādi. Otrs papildināts izdevums* [Latvian Medicinal Plants. Second supplemented edition], Riga.
16. Ilsters 1883, 1884, 1885 (see note 8 above).
17. Karlsone, A., Korsaka, I, and Siliņa-Piņķe, R. (in press) *Krāsošana ar augiem. Vācbaltiete Marta Bīlenšteina un latviešu nemateriālais kultūras mantojums* [Dyeing with Plants. The Baltic German Martha Bielenstein and the Intangible Cultural Heritage of the Latvians], Riga.
18. These records are stored in the Repository of Ethnographic Material of the Institute of Latvian History, University of Latvia.
19. Madre, I. (1990) *Krāsošana ar augu krāsvielām* [Colouring with Plant Dyestuffs], Riga.

Revealing the Mysteries of the Madder Bath

Vincent Daniels

Abstract The aqueous extract of madder root contains several hydroxyanthraquinone components which are responsible for its dyeing and pigment-making properties. The concentrations of the individual compounds in the madder bath depend not only on the initial concentrations in the roots, but also on their solubilities as the maximum concentration in the extract is dependent on the latter. There is very little data in the literature on the solubilities and this study has determined the solubility characteristics of alizarin, purpurin and pseudopurpurin over a range of temperatures in both water and 5% potassium aluminium sulfate solution. It is anticipated that the results will aid in elucidating the way madder dye and pigment baths operate. The solubilities all increase significantly with temperature and the solubilities of pseudopurpurin in both water and 5% alum are much higher than previously reported in the literature. Pseudopurpurin is not available commercially and a method for isolating it has been developed which utilises the solubility data derived in this work. Alizarin has a very low solubility in water at room temperature but it was found that, in a pigment made from a madder extract at room temperature, the amount of alizarin relative to pseudopurpurin at room temperature was higher than predicted from its solubility in water. The reasons for this apparent anomaly are discussed. The stability of pseudopurpurin was examined in both a water and alum solution, also in a heated solution of methanol and 10% sulfuric acid.

Introduction

The root of the madder plant *Rubia tinctorum* L. has been used for textile dyeing and for the manufacture of pigments for thousands of years.[1] There are many methods for textile dyeing and for pigment making that employ the extraction of water-soluble compounds from the madder root into aqueous liquids, often pure water, alum or alkali. The hydroxyanthraquinones (HAQs) are present both in the free form (aglycones) and with a sugar group attached (glucosides or glycosides), which imparts greater water solubility. When dyed textiles or madder-derived pigments are analysed the compounds detected are the HAQs, which either dye the textile fibre directly or by way of a metal mordant. Several methods for pigment making adsorb the HAQs onto freshly precipitated hydrated alumina, which may be made by adding a solution of potassium aluminium sulfate (alum) to an alkaline solution (e.g. potassium carbonate) or a basic solid (e.g. calcium carbonate). When madder-derived pigments are analysed, the components frequently found are alizarin and purpurin together with pseudopurpurin. However, the method of analysis is crucial to the result as the extraction process may convert pseudopurpurin to purpurin by a process of decarboxylation that is accelerated by acids and heat. Methods of analysis that preserve pseudopurpurin and other labile components are of recent discovery and so a review of past analyses must consider this complication.[2] Recent analyses that use 'soft extraction' often find glycosides of alizarin and lucidin and additional aglycones such as lucidin, nordamnacanthol and rubiadin,[3] however pseudopurpurin can be completely missed in liquid chromatography–mass spectrometry (LC–MS) analysis due to ionisation problems.[4]

Analysis of laboratory-made pigments produced from water extracts of madder often reveals the presence of approximately equal amounts of alizarin and purpurin plus pseudopurpurin. However, a survey of pigment analyses found in the literature showed that pink pigments from antiquity often contained low levels of alizarin, but high levels of purpurin or pseudopurpurin. Previous work from this laboratory also discovered that madder pigments of the characteristic compositions found from antiquity could be obtained by extracting the madder root directly into an alum solution instead of water; these extracts, and the pigments made from them, had low concentrations of alizarin with high concentrations of pseudopurpurin.[5] The cause of this was the relatively high solubility of pseudopurpurin in alum compared with that of alizarin. It became apparent that a systematic study of the solubilities of the principal madder components in both water and alum might be of great value when trying to understand the behaviour of the madder pigment bath.

Very little data seems to exist on the solubility of madder components in water or alum. The solubility of alizarin in water has been found to be at 18 °C, 0.504 mg L^{-1}, at 25 °C, 0.60 mg L^{-1} and at 100 °C, 3300 mg L^{-1}.[6] According to Schweppe and Winter, alizarin is 'very sparingly soluble' at 20 °C, and at 100 °C the solubility is 300 mg L^{-1}.[7] Purpurin is said to be 'soluble in alum solution and sparingly soluble in hot water', while it is claimed that pseudopurpurin is 'insoluble in cold water' and 'dissolves slightly in alum solution but simultaneously gives rise to an insoluble alumina lake'.

Consideration of the molecular formulae for the HAQs reveals that they have differing numbers of hydroxyl groups: alizarin has two, purpurin three and pseudopurpurin four. The solubility of the compounds ought to increase with increasing numbers of hydroxyl groups and the corresponding hydrophilic nature of the compounds, thus it is surprising that Schweppe and Winter should describe pseudopurpurin as the *least* soluble of the principal three HAQs when it might be predicted that it should be the *most* soluble. The statement on the solubility of pseudopurpurin in alum was also worthy of further investigation in view of its ambiguity and the high solubility observed in previous work in this laboratory.

In view of the low solubility of alizarin at ambient temperatures, it is worth considering whether typical recipes for making madder pigments use sufficient water to enable all the alizarin present to dissolve. De Santis and Moresi determined the amount of alizarin which can be leached out of madder root at 25 °C over 12 hours.[8] They found 0.021% of free alizarin while the glycoside form, ruberythric acid, was present at 0.26%, giving a total of 0.28% overall. Variation of alizarin content is to be expected in madder and indeed other researchers find higher and lower alizarin contents;[9] however, it is instructive to do a rough calculation to determine how much alizarin might be extracted from the madder in a typical pigment-making process. Derksen *et al.* found that their madder sample initially contained ruberythric acid, but this hydrolysed to alizarin over a 24-hour period merely by allowing it to stand in water.[10] If this were to be the case during a process in which the madder is soaked for a similar period, the *total* alizarin figure above is the one to use to assess how much alizarin would be available.

Most pigment preparations are performed at elevated temperatures, however if a preparation was carried out at 25 °C, a litre of water extract should contain a maximum of only 0.60 mg of alizarin due to the solubility limit. If madder contained the 0.28% alizarin found by De Santis and Moresi, it can be calculated that a maximum of 2.14 g L^{-1} of madder could be used before incomplete extraction became theoretically possible and the composition of the pigment would be controlled by the solubilities of the HAQs rather than the ratio of HAQs in the madder itself.

To take one example of a pigment recipe, the Standard method (code ML-std) reported by Kirby *et al.* uses 2 g of madder extracted by 60 ml of water at 70 °C,[11] which corresponds to 33.33 g of madder in a litre. Although at elevated temperatures all the alizarin might be extracted by water, at a lower temperature such as 20 or 25 °C, all the alizarin cannot be removed by that quantity of water. Henderson has found a water/ethanol extract of madder to contain principally polysaccharides, free sugars and tannins.[12] Although the solubilities of these HAQs are important basic data, it must be remembered that the solubility of HAQs in a madder extract actually depends on their solubility in an aqueous solution of these carbohydrates and tannins rather than in pure water.

Materials

Madder root from an undisclosed Middle Eastern location, purchased from L. Cornelissen & Son in 2012, was reduced to powder in a coffee grinder prior to use. Alizarin was purified from (brown) Aldrich alizarin by dissolving it in toluene and allowing the filtered orange solution to evaporate; orange crystals were obtained while a brown powder was observed on the filter paper. Purpurin (Aldrich) was used without purification. Pseudopurpurin was donated for these experiments by David Hill who inherited a sample made in approximately 1936 by his father, the eminent dye chemist Robin Hill. This sample was shown to be pure by proton nuclear magnetic resonance (NMR) at 500 MHz.[13] Solubilities of compounds of low solubility are difficult to determine gravimetrically because the amounts of solute are so small, so a method of comparing solubilities by the transmission UV/visible spectra was adopted. It was assumed that the Beer-Lambert law was applicable, that is, the absorbance (optical density) was proportional to the concentration of the solute.

Experimental

Solubility of madder

The solubility of madder in water was obtained at room temperature (20 °C) by soaking 1 g of powdered madder for 24 hours in 100 ml of distilled water. The extract, which was separated from the madder by filtering, evaporated to dryness at 100 °C. After conditioning in ambient conditions, the mass of the residue was determined and the solubility calculated. The solubility of madder at 60 °C was determined by storing the madder/water mixture at room temperature for 24 hours, then at 60 °C for an hour before filtering while still hot. The procedure was then as for the room temperature sample.

Solubility of hydroxyanthraquinones

Saturated solutions of the solutes were obtained by adding an excess of the studied solute to water or to 5% w/w alum (aluminium potassium sulfate) solution in a sealed container. The contents were shaken and left for 7 days at the temperature of interest. At the end of this period a sample of the supernatant liquid was transferred to a 10 mm path length quartz cuvette so that its UV/visible spectrum could be determined in the range 300–800 nm. In some cases warm solutions were diluted with the appropriate solvent before measurement to avoid precipitation of the solute. UV/visible spectrometers cannot reliably measure absorbances of greater than 2.0 so if the solutions were too concentrated they were diluted before measurement was performed to bring the absorbance into the 0.5–2.0 range. The results for diluted samples were multiplied by the dilution factor to find the 'calculated absorbance' of the solution.

Graphs were plotted of the absorbance of the solutions against temperature (Figs 1 and 2). In the 'alizarin in water' graph (Fig. 1a), the data points derived from literature values of solubility are included (see Results section below). Experimental error in all the graphs is estimated to be ± 10%.

Determination of extinction coefficients

To obtain approximate correlations between absorbance measurements and concentration in solution, the extinction coefficient of some of the HAQs in water were obtained experimentally. The procedure was to obtain a saturated solution at 60 °C and then measure its absorbance. The same solution was allowed to evaporate in a weighed beaker and the amount of solute residue subsequently calculated. A blank determination with only solvent was performed, however no detectable residue was found. The equation absorbance = εcl was used to calculate ε (the extinction coefficient), where c is the molar concentration and l the path length in cm. The errors on the extinction coefficient are estimated to be ± 20% and are relatively large due to the error in weighing very small amounts of non-volatile residues.

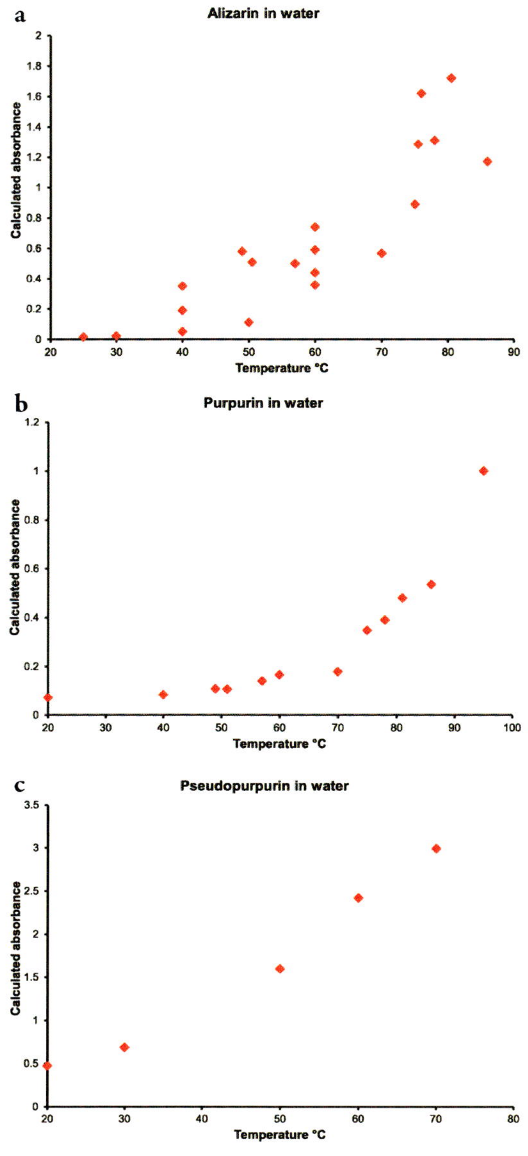

Figure 1 The change in solubility of (a) alizarin, (b) purpurin and (c) pseudopurpurin in water with temperature, as indicated by their absorption in saturated solutions.

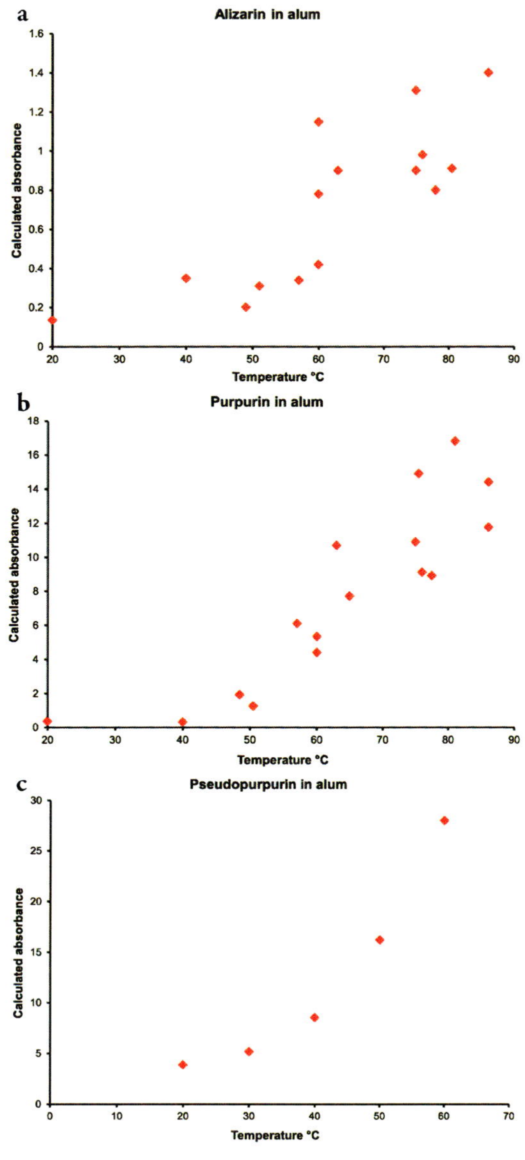

Figure 2 The change in solubility of (a) alizarin, (b) purpurin and (c) pseudopurpurin in 5% alum solution with temperature, as indicated by their absorption in saturated solutions.

Method for isolating pseudopurpurin from madder root

The following method for isolating pseudopurpurin from madder root utilises the solubility properties of pseudopurpurin in alum. The principles behind the extraction are as follows: the initial stage is the extraction of pseudopurpurin from the madder into the alum solution in which it is very soluble; however some alizarin is also extracted. The crude pseudopurpurin/alum solution is freed of alizarin by shaking it with toluene: the alizarin goes into the toluene while the pseudopurpurin remains in the aqueous phase. The aqueous layer is next acidified and the pseudopurpurin then dissolves in the toluene. The quantities of materials used are not critical, but suggested quantities are given in brackets for those attempting this for the first time.

Madder powder (5 g) is mixed with water (25 ml) and left for 24 hours. It is then filtered onto a Büchner funnel and the brown filtrate discarded. The residue is added to 5% alum in water (250 ml), the mixture is shaken and heated to 50 °C for approximately 30 minutes. The red solution is then filtered free of madder particles and kept. The red solution is added to toluene (100 ml), which is immiscible and forms an upper layer. After several days the toluene layer becomes yellow or orange due to the presence of alizarin and also purpurin if present. The toluene layer is discarded and 2.5M sulfuric acid (10 ml) added. A further 100 ml of toluene is added and the mixture shaken and left for several days. The pseudopurpurin migrates into the toluene layer which is isolated and the toluene allowed to evaporate, leaving a residue of red pseudopurpurin.

Pigment making

To test the validity of the solubility measurements and validate their potential usefulness, two pigments were prepared from a solution saturated in both alizarin and pseudopurpurin. One pigment was made at 20 °C and the other at 60 °C. Pigments were made by taking 100 ml of the saturated solution and adding it to 50 ml of 5% alum solution. This was warmed to 60 °C and poured into 50 ml of water in which was dissolved 0.094 g of potassium carbonate. The resulting pigment was filtered off and dried. The pigment was analysed by the analytical scheme described by Hill and Richter.[14] A quantity of pigment was placed in a screwcap container and 5 ml of toluene and 5 ml of 2.5M sulfuric acid added. The contents were shaken and left for at least 24 hours when the coloured, immiscible toluene layer was extracted and examined by UV/visible spectroscopy. Trends in the relative concentrations of alizarin and pseudopurpurin are indicated by comparing the peak heights at the wavelength absorption maxima for alizarin (243 nm) and pseudopurpurin (491 nm); pseudopurpurin also has smaller absorption maxima at 459 nm, 528 nm and 560 nm. Pigments were also made from madder powder using the ML-std method at 20 °C and 60 °C and analysed in the same way.

Stability of pseudopurpurin

Pseudopurpurin can decarboxylate with the formation of purpurin.[15] This phenomenon is most frequently reported during the acid hydrolysis of dyed textile or pigment samples prior to dye analysis. Solutions of pseudopurpurin in water and 5% alum were stored at ambient temperature or in a thermostatted oven to determine how long it would take before purpurin could be seen in the solution. Analysis of the stored solutions was conducted using the Hill and Richter method by acidifying the solution and extracting the HAQs into toluene where any purpurin formed could be detected by its characteristic absorption maximum (479 nm) in the UV/visible spectrum.

Results

Solubility of madder

The mass of madder root extracted into water (1 g dispersed in 100 ml of water) at 20 °C was found to be 58% and at 60 °C, 60%. Experimental error was ± 1%.

Determination of extinction coefficients

Table 1 shows the experimentally obtained extinction coefficients in water and the values listed from R.H. Thomson, *Naturally Occurring Quinones*, for the same solutes but in ethanol.[16] The wavelength for measurement was that of greatest absorbance in the 400–500 nm region.

Solubility of hydroxyanthraquinones

The graphs for the comparative solubility of the HAQs are shown in Figure 1. In the graph for alizarin (Fig. 1a), the points at 20 and 25 °C are taken from the Merck Index,[17] by taking the solubility values and calculating the absorbance in a 10 mm cell using the extinction coefficient determined experimentally in this work.

Table 1 Visible absorption maxima and extinction coefficients (ε) of HAQs in water and ethanol.

	In water		In ethanol	
Solute	Wavelength (nm)	ε	Wavelength (nm)	ε
Alizarin	438	5950	434	5010
Purpurin	484	12815	485	8320
Pseudopurpurin	487	10601	487	9120

Method for isolating pseudopurpurin from madder root

The madder sample used seemed to be free of purpurin as this was never seen in the toluene extracts made from the roots. In one experiment the yield of pseudopurpurin was 0.12%. The UV/visible spectrum of the product dissolved in toluene did not reveal any other peaks, for example from alizarin or purpurin. Experiments by Hill and Richter on pseudopurpurin showed the pseudopurpurin content of their sample of madder was 0.17%.[18]

Pigment making

Pigments were prepared successfully from both solutions. After the analytical procedure the two spectra obtained (Fig. 3) show that there is no detectable alizarin in the pigment made from the saturated solutions at 20 °C, but approximately equal amounts of alizarin and pseudopurpurin at 60 °C. The solubility graphs confirm that while alizarin is practically insoluble at 20 °C, its solubility is about 35 times higher at 60 °C. Pseudopurpurin is also more soluble at 60 °C, but by a factor of about 5.3. Pigments prepared from madder powder showed that there were approximately equal amounts of alizarin and pseudopurpurin in the pigments made at both 20 °C and 60 °C.

Stability of pseudopurpurin

The pseudopurpurin sample was approximately 80 years old and no degradation products were found by proton NMR, thus it can be concluded that solid pseudopurpurin is stable when pure and if stored in ambient conditions in the dark. In aqueous solution at 20 °C, no purpurin was seen after 11 weeks

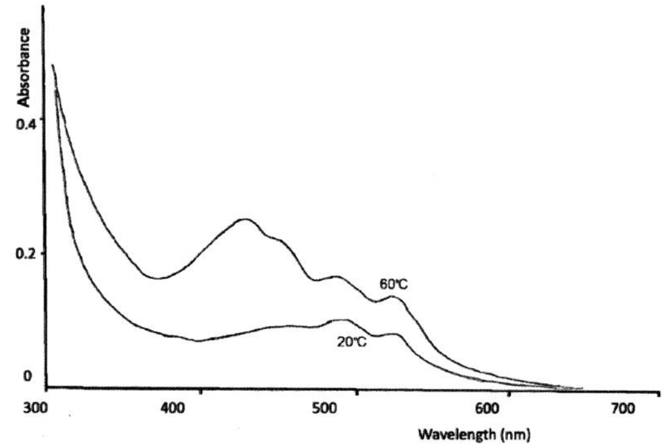

Figure 3 Spectra of the toluene extracts from acidified pigments made from a saturated solution containing both alizarin and pseudopurpurin at 20 °C and 60 °C.

storage; however, at the next measurement at 14 weeks storage a large amount of degradation had taken place. At 45 °C, purpurin was detected after 7 days. Dissolved in 5% alum, there was no purpurin detected after 16 weeks at ambient temperature or after 2 weeks at 70 °C, but complete conversion to purpurin was detected after 4 weeks at the latter temperature.

Conclusion

The solubilities of alizarin, purpurin and pseudopurpurin have been determined over a range of temperatures for two solvents, water and 5% alum in water. The solutes all increased their solubilities substantially as the temperatures increased. Of particular note was the good solubility of pseudopurpurin in both an alum solution and water, which was contrary to the statement by Schweppe and Winter that pseudopurpurin is insoluble in water and forms an insoluble aluminium compound in alum. Of the three compounds, pseudopurpurin had the greatest solubility in water, consistent with the compound being the most hydrophilic as indicated by the greatest number of hydroxyl groups. The solubility of pseudopurpurin in alum was so high that it could be used to devise a method for isolating pseudopurpurin from madder root, which gave a yield comparable to one value quoted for the pseudopurpurin content of madder root. The solubility of alizarin is particularly low in water at 20 °C, but increases rapidly with increasing temperature.

Pigments were prepared at 20 and 60 °C from a solution containing saturated amounts of alizarin and pseudopurpurin. Analyses showed that the pigment made at the lower temperature had relatively very little alizarin, while the pigment made at higher temperature had relatively more alizarin than pseudopurpurin as indicated by the peak heights. This was in accordance with the solubility results, which indicate that alizarin's solubility increases rapidly with temperature. The solubility/temperature graphs confirm that when increasing the temperature from 20 °C to 60 °C, the solubility of alizarin increases by approximately 35 times, while that of pseudopurpurin increases fivefold. However, when the experiment was repeated with madder powder, the peak heights for alizarin and pseudopurpurin were similar at both temperatures. This indicates that the alizarin may be present in a different form in the water extract of madder powder.

Recent research has suggested that glycosides play a much greater part in dyeing processes than hitherto expected in that glycosides of alizarin are present in much higher concentrations than alizarin in madder extracts, and that the glycosides also have a higher affinity for wool than alizarin.[19] There are also interactions between adsorbed species which suggest aggregation of alizarin-containing molecules on the wool. If this hypothesis could be extended to pigments made from madder, the experimental results could be explained as the glycosides (and glucosides if present) should have good solubility over a wide range of temperatures. However, opposing this view is our experience that ruberythric acid does not easily form pigments or dye aluminium-mordanted wool, and even then the products are yellow as opposed to the expected red, brown or orange produced with alizarin. An alternative hypothesis might be that the alizarin is not being dissolved in pure water but in a complex mixture of natural products that dissolve out of the madder root and which is a better solvent for alizarin than water. The experimental results showed that madder powder (1 g in 100 ml of water) dissolved up to 60% of its initial mass in water. The standard madder pigment method quoted uses a ratio of 33.33 g of madder dispersed in a litre of water. If the 60% solubility were maintained at the higher madder/water ratio used in the pigment preparation, the amount of dissolved material in water would be 20 g L^{-1} of dissolved polysaccharides and tannins, which might be a better solvent for alizarin than water.

Recent research is changing our view of what occurs in the madder bath while dyeing and pigment making. It is possible that pseudopurpurin may be degraded to purpurin during the preparation of madder root for sale. However, the sample used in this research contained pseudopurpurin and no purpurin detectable by the Hill and Richter method used for analysis. The present research has shown that when initially present in the madder sample, pseudopurpurin has a significant presence in both pure water and in alum extracts. Alizarin has a very low solubility in water at 20 °C, but its substantial presence in pigments made from madder at that temperature suggests that the alizarin may be taken up from alizarin glycosides/glucosides or alizarin in association with some organic moiety.

Solid pseudopurpurin is very stable, but in neutral aqueous solutions it will decarboxylate to purpurin over several weeks. These observations suggest that pseudopurpurin extracted from madder in heated containers probably converts to purpurin over several hours or days. Other transformations occurring, such as the hydrolysis of ruberythric acid to alizarin and oxidation of lucidin to nordamnacanthol, combine to make the madder bath a complicated changing system. From the few experiments carried out to date it can be seen that solutions in alum appear to be more stable than solutions in water.

Editor's note and author's address

Address for correspondence: Department of Scientific Research, The British Museum, London WC1B 3DG; vdaniels@britishmuseum.org; daniels2@ntlworld.com. Paper received by 13 March 2018; revised version received by 16 September 2018. This paper was presented at *Dyes in History and Archaeology* 36, Hampton Court, UK, 26–28 October 2017.

Acknowledgements

The author is grateful to Dr David Hill for donating the pseudopurpurin sample, to Dr Lauren Hill previously of Leeds University for analysing the pseudopurpurin, also to the Department of Scientific Research at the British Museum for providing the facilities for this research.

Notes and references

1. Cardon, D. (2007) *Natural Dyes: Sources, Tradition, Technology and Science*, London.
2. Daniels, V., Devièse, T., Hacke, M. and Higgitt, C. (2014) 'Technological insights into madder pigment production in antiquity', *British Museum Technical Researh Bulletin* 8, pp. 13–28.
3. Ford, L., Henderson, R.L., Rayner, C.M. and Blackburn, R.S. (2017) 'Mild extraction techniques using aqueous glucose solution for the analysis of natural dyes in textile artefacts dyed with dyer's madder (*Rubia tinctorum* L.)', *Journal of Chromatography A* 1487, pp. 36–46.
4. Dr Lauren Ford, personal communication, 2018.
5. Daniels *et al.* 2014 (see note 2 above).
6. *The Merck Index: An Encyclopedia of Chemicals and Drugs* (1968) 8th edn, ed. P.G. Stecher, Rahway, NJ.
7. Schweppe, H. and Winter, J. (1997) 'Madder and alizarin', in E.W. FitzHugh (ed.), *Artists' Pigments: A Handbook of their History and Characteristics*, vol. 3, Washington, National Gallery of Art; London, Archetype Publications, pp. 109–42.
8. De Santis, D. and Moresi, M. (2007) 'Production of alizarin extracts from *Rubia tinctorum* and assessment of their dyeing properties', *Industrial Crops and Products* 26 pp. 151–62.
9. Angelini, L.G., Pistelli, L., Belloni, A. and Panconesi, S. (1997) '*Rubia tinctorum* a source of natural dyes: agronomic evaluation, quantitative analysis of alizarin and industrial assays', *Industrial Crops and Products* 6, pp. 303–11; Lhodi, A.H., Sant'Ana, A.E.G. and Charlwood, B.V. (1994) 'Quantitative analysis of alizarin in tissue cultures of *Rubia* species by high performance liquid chromatography', *Phytochemical Analysis* 5, pp. 261–5.
10. Derksen, G.C.H., Naayer, M., van Beek, T.A., Capelle, A., Haaksman, I.K., van Doren, H.A. and de Groot, Æ.(2003) 'Chemical and enzymatic hydrolysis of anthraquinone glycosides from madder roots', *Phytochemical Analysis* 14, pp. 137–44.
11. Kirby, J., van Bommel, M. and Verhecken, A. with Spring, M., Vanden Berghe, I., Stege, H. and Richter, M. (2014) *Natural Colorants for Dyeing and Lake Pigments: Practical Recipes and their Historical Sources*, London, Archetype Publications in association with CHARISMA, p. 91.
12. Henderson, R.L. (2013) 'The chemical profile of *Rubia tinctorum* in wool dyeing and a novel fibre extraction method for compositional analysis' (PhD thesis, University of Leeds).
13. Ford 2018 (see note 4 above).
14. Daniels *et al.* 2014 (see note 2 above); Hill, R. and Richter, D. (1937) 'Anthraquinone pigments in *Galium*', *Proceedings of the Royal Society B* 121, pp. 547–60.
15. Wouters, J. (2001) 'The dye of *Rubia peregrina*: I. Preliminary investigations', *Dyes in History and Archaeology* 16/17 pp. 145–57.
16. Thomson, R.H. (1971) *Naturally Occurring Quinones*, 2nd edn, London.
17. *The Merck Index* 1968 (see note 6 above).
18. Hill and Richter 1937 (see note 14 above).
19. Ford, L., Rayner, C.M. and Blackburn. R.S. (2017) 'Comparative sorption isotherms for colorants present in dyers' madder (*Rubia tinctorum* L.) provide new insights into historical dyeing', *Coloration Technology* 134, 1, pp. 1–10.

Analysis of Red Yarns in Textiles from the Huaca Malena Archaeological Site in Peru

Laura Maccarelli,[1*] Terry T. Schaeffer,[1] Elizabeth Burr[2]
and Ioanna Kakoulli[3]

Abstract Textiles recovered from the Huaca Malena archaeological site in the Asia Valley on the south-central coast of Peru are representative of several pre-Columbian cultures. They display a variety of iconographies and a range of techniques. Wari influence at the site has been surmised on the basis of motifs in the artefacts and the burial customs. Early pre-Colombian coastal cultures used *Relbunium* spp. for red dyes, but the Wari are known to have used cochineal. Thus it was of interest to determine which red dyes were present in the Huaca Malena textiles that stylistically represented the various pre-Colombian cultures. Red and pink yarn samples were taken from seven textiles recovered from the Huaca Malena mausoleum representing Wari and local Malena styles. Yarn spin and fibre morphology were assessed by polarised light microscopy and scanning electron microscopy. Dyes were analysed by high-performance liquid chromatography with diode array detection, fibre optic reflectance spectroscopy and surface enhanced Raman spectroscopy. The archaeological yarns analysed were all of animal (camelid) origin tinted with cochineal. This result is consistent with the Wari dyeing custom and suggests Wari influence on the other cultures.

Introduction

In pre-Colombian Peru during the Middle Horizon period (*c*.600–1000 CE) several cultures coexisted (e.g. Wari, Tiwanaku and Nazca). One of these cultures was the Wari, whose capital was located in the south-central Andes near the modern city of Ayacucho, Peru. During this period the Wari expanded from their homeland to interact with, and eventually dominate, other groups over a wide area from the south-central Peruvian coast to the Andes north of present-day Lima.[1] The extent of Wari domination and the nature of their interactions with other native cultures continue to be revealed by recent archaeological excavations.[2]

Peru is divided into three geologically diverse regions by the Andean mountain chain: high elevation, mid-low elevation and coastal areas. In the high elevation region, where the weather is cold, domesticated camelids such as alpaca and llama were raised not only for their meat but also for their wool, which was used in textile production.[3] In this region, *Dactylopius coccus* insects, the source of cochineal dye, have also been cultivated on *Opuntia* spp. cacti for more than a thousand years.[4] At mid-low elevations, the climate is more moderate and agriculture is possible due to the annual rains. The coastal areas have arid river valleys; here the local cultures relied on seafood for protein and grew cotton for textile production.[5] *Relbunium* spp. are native to Peru's tropical and temperate biomes at mid-to-high elevations, as well as the lomas regions along the desert coast of Peru.[6]

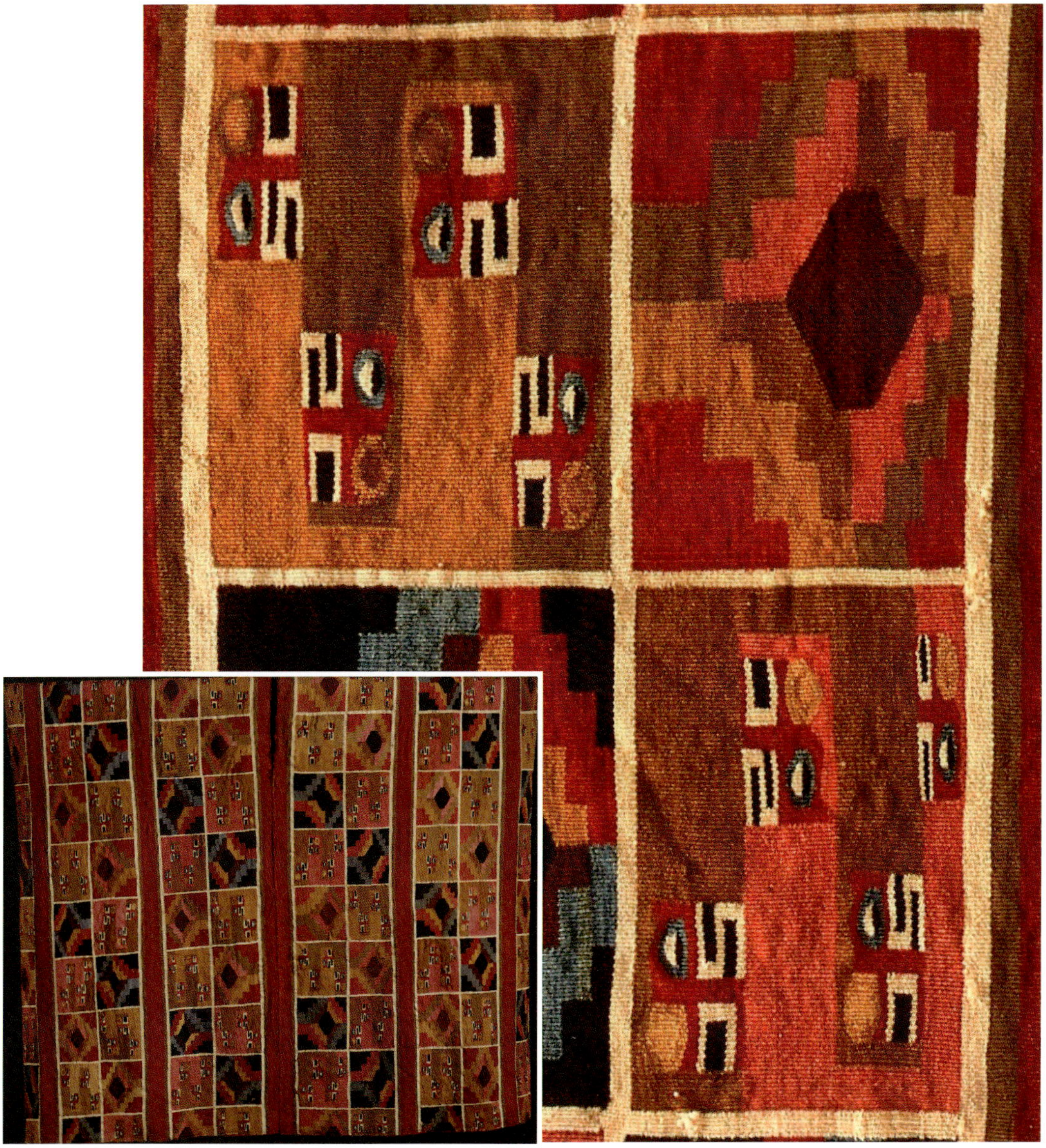

Figure 1 Detail of a Wari man's tunic showing how the differently coloured design elements combine to form geometrical patterns. Camelid fibre and cotton, 96.5 × 110.5 cm. Los Angeles County Museum of Art, 70.3.1. (Photo © Museum Associates/LACMA)

Textiles occupied a most important place in Wari culture – not only were they very highly valued, they were also bestowed on occasion as political favours, and appear to have had great symbolic significance.[7] They were mostly produced using a sophisticated tapestry weave, and they bore highly stylised designs in several colours arranged to create overall geometric patterns (Fig. 1). The imagery in the textiles was formed with camelid yarns, frequently on a cotton ground. The use of cotton suggests a trade or tribute system in which the Wari interacted with coastal cultures to obtain the plant fibre.[8] In addition, stylistic evidence suggests that as the Wari expanded into the territories of other cultures, they also disseminated their manufacturing techniques and iconography. Consequently,

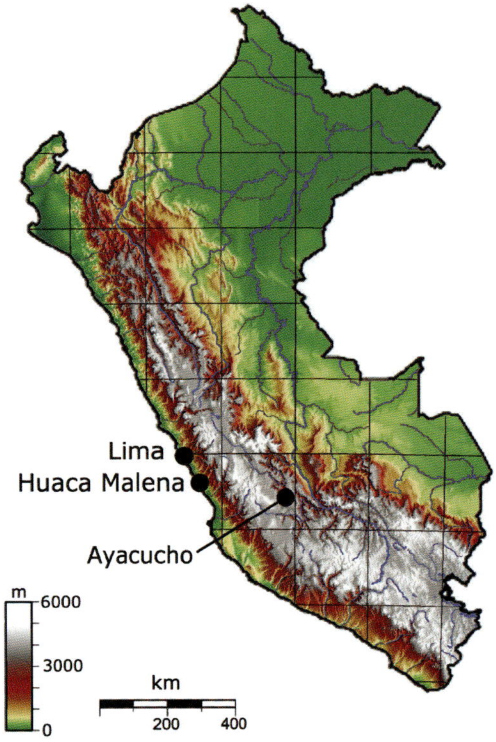

Figure 2 Map of Peru indicating the locations of the Wari capital near Ayacucho and Huaca Malena on the coast south of Lima. (Map © Elizabeth Burr)

textiles with a mixture of Wari and local features were produced.

Huaca Malena is located in the arid Asia Valley about 110 km south of Lima on the south-central coast of Peru (Fig. 2). The site consists of a 4-acre artificial platform on top of which sit six smaller terraces with walls of handmade, semi-cylindrical adobe bricks. Ceramics recovered from this site suggest it was built about 200 CE in the Early Intermediate period and was occupied until around 550 CE.[9] The structures may have served a variety of purposes during this time. At some point in the Middle Horizon after the site had been abandoned, it appears to have been repurposed by the Wari as a mausoleum, and intrusive tombs containing multiple mummy bundles were dug into the structures.[10]

The site of Huaca Malena was first excavated by Tello and Mejía Xesspe in 1925.[11] During the first campaign of their excavations, they recovered over 300 funerary bundles from the upper platforms. Soon after, the site was abandoned and a long period of indiscriminate looting occurred. In 1997, in a second excavation campaign led by Denise Pozzi-Escot and Rommel Ángeles Falcón, about 50 more intact funerary bundles and a large number of fine textiles and fragments were found.[12] Among the textiles recovered, several styles have been recognised including Wari, a local style identified as the Malena style, and examples of mixed technique.[13] These objects could suggest some reciprocal interaction among the cultures, rather than total domination by the Wari.

The recovery of textiles from the Huaca Malena site that appear stylistically to belong to different cultures (Malena and/or Wari) provides an opportunity to characterise differences between their textile techniques. One variant that has been observed by several researchers is the use of red organic dyes from either a plant or insect source. Whereas cochineal has been found almost exclusively in Wari and later textiles, anthraquinones typical of *Relbunium* spp. plants have been identified in a large majority of yarns from objects dated to Early Horizon and earlier cultures.[14] Because some of the Huaca Malena items appear to be in a mixed style, incorporating both Malena (a coastal style) and Wari elements, the question becomes specific: had Wari influence in the Asia Valley imposed the use of the insect dye, while local designs continued to be incorporated into textiles? The identification of the red dyes in yarn samples from seven archaeological textiles from Huaca Malena was undertaken to address this question. Preliminary analyses of these archaeological samples suggested the use of cochineal.[15]

Materials

Reference materials

A reference yarn sample dyed with *Relbunium hypocarpium* (L.) Hemsl. (#16) by Max Saltzman (now in the collection of the UCLA/Getty Conservation Program) was used as a reference standard for *Relbunium* dyestuff, which contains purpurin and pseudopurpurin but not alizarin.[16] This plant is indigenous to South America and was used by local cultures at least as early as the beginning of the Middle Horizon period.[17] As standards for cochineal, a solution of approximately 2 mM cochineal from Aldrich

Table 1 Archaeological samples. (Photos of textiles © Elizabeth Burr)

Sample name	Image	Style	Description
VM01	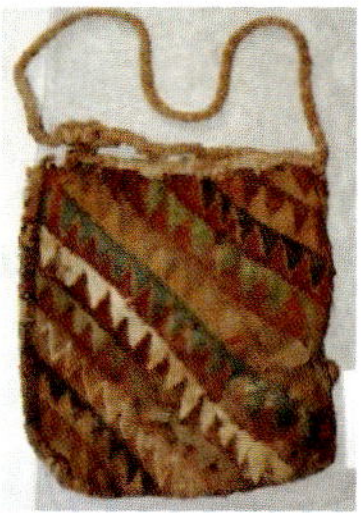	Mountain style	Bag, eccentric tapestry technique, camelid warp and weft
T-130A	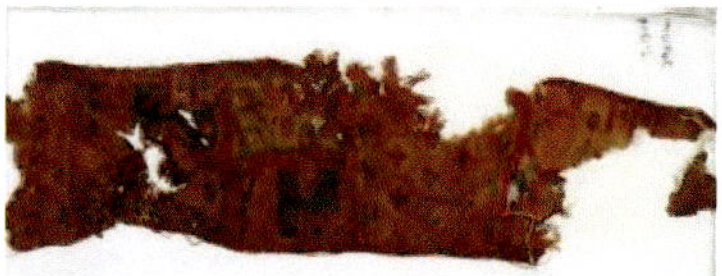	Wari style	Band, eccentric tapestry technique, cotton warp and camelid weft
VM04	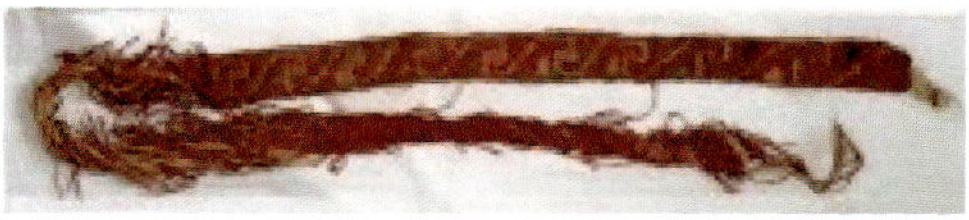	Malena style	Polychrome band, double weave technique, undyed cotton and dyed camelid
061.02.01	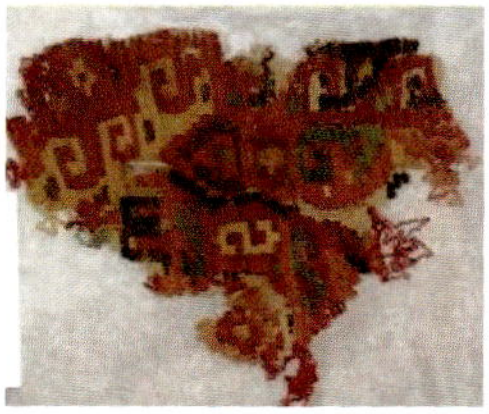	Wari/Malena style	Textile fragment, tapestry technique, dyed camelid
T-69	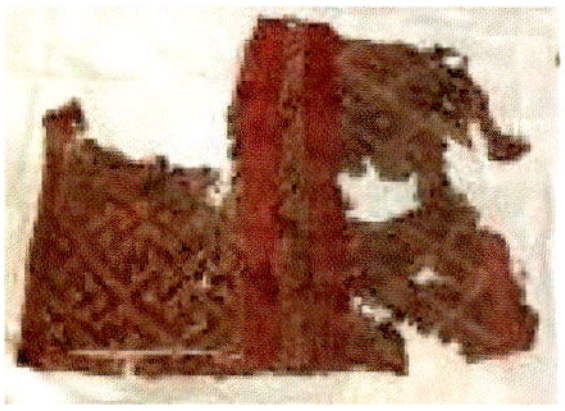	Malena style	Cloth, slotted tapestry technique, cotton warp camelid weft
77.10.02	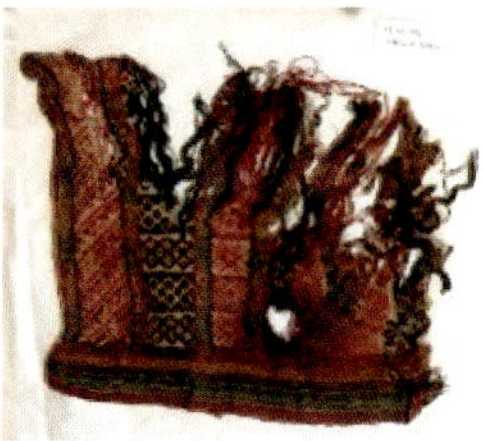	Malena style	Corner of cloth, brocade technique, cotton and camelid
119.000.029	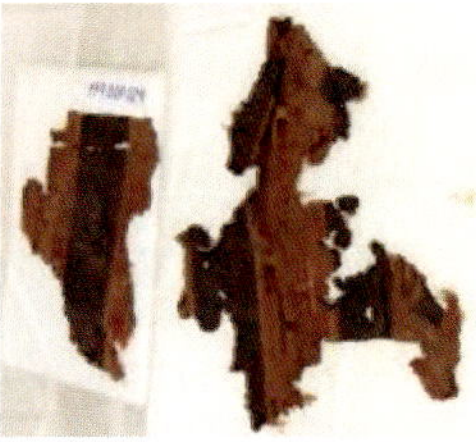	Wari style	*Uncu* fragment, eccentric tapestry technique, camelid warp and weft

prepared in ethanol water (3:1, v/v), and a cochineal-dyed woollen yarn sample from the Helmut Schweppe reference collection in the Getty Conservation Institute (Sch #527, cochineal 50% on wool, mordanted with 25% alum and 6% potassium bitartrate, also commonly known as cream of tartar) were used.

Archaeological samples

Red-dyed yarn samples were removed from seven archaeological textiles that had been recovered from the Huaca Malena site by Pozzi-Escot and Ángeles Falcón. These textiles are stored at the Museo Municipal Huaca Malena in Capilla de Asia, Peru. Samples of weft yarn were taken from a bag, two bands and four textile fragments with a variety of styles. The samples were identified by Ángeles Falcón as follows: two as Wari style, three as the local Malena style, and one as a mix of Wari and Malena styles.[18] The bag was considered to represent a mountain style. The features of these textiles are summarised in Table 1.

Chemicals

Acetonitrile (HPLC grade), methanol (HPLC grade), water (HPLC grade) and hydrochloric acid (36.5–38% w/w, Certified ACS Plus) were obtained from Fisher Scientific, formic acid (LC-MS grade) from Thermo Scientific and Drierite was obtained from W.A. Hammond Drierite Company, Ltd. Denatured anhydrous ethanol (90% ethanol, 5% methanol, and 5% isopropanol, HPLC Grade) were obtained from Alfa Aesar; potassium nitrate (99% ACS reagent) from Acros organics; silver nitrate (99.9999%) from Aldrich Chemical Co; and hydrofluoric acid (48 wt%), sodium citrate tribasic dihydrate (99% ACS reagent) and citric acid (99.5% ACS reagent) from Sigma Aldrich.

Methods

Microscopy

The archaeological samples were rinsed in ethanol to remove debris that might have remained after the excavation process. Individual fibres from each archaeological sample were mounted for scanning electron microscopy on an SEM stub using carbon tape, and coated with gold using an Anatech Hummer 6.2 sputtering system under vacuum. A FEI Nova NanoSEM230 variable pressure (VP) scanning electron microscope was used at high vacuum with a secondary electron detector.[19] Planar optical microscopy and transmitted microscopy were performed with a Keyence VHX-1000 digital optical microscope and an Olympus BX51 polarised light microscope with an attached Nikon D90 SLR camera, respectively. A Nikon Microphot polarised light microscope in transmitted mode with a Leica DMC4500 camera was also used for fibre documentation. Micrographs of the fibres were measured using ImageJ to acquire accurate diameters. Fifteen different individual fibres from each archaeological sample were measured and averaged to increase the accuracy of the results.[20]

Fibre optic reflectance spectroscopy (FORS)

FORS was conducted on reference and archaeological fibres using a Varian Cary 50 UV-Vis spectrophotometer equipped with an optical fibre and a Harrick Scientific Barrelino accessory. The fibre samples were placed on a white non-fluorescent paper background (Whatman 1 paper). The Barrelino camera permitted accurate positioning of the accessory over the fibres to optimise the reflectance spectral intensity. The reflectance data were smoothed, converted to the Kubelka-Munk transforms, and the second derivatives calculated by the WinUV software supplied with the Cary 50 instrument. The accuracy of the band positions in the resulting spectra is ± 3 nm.

High-performance liquid chromatography with diode array detection (HPLC–DAD)

For HPLC–DAD analysis, the fibres from each yarn sample were treated with 100 μL of a mixture of 37% hydrochloric acid in water and methanol (2:1:1, v/v/v) at *c.*100 °C for 10 min.[21] The extracts were dried in a desiccator under vacuum. The residues were dissolved in 60 μL of a mixture of water/methanol (1:1, v/v) and the resulting solutions centrifuged for 5–8 min at 12,100 × *g*.

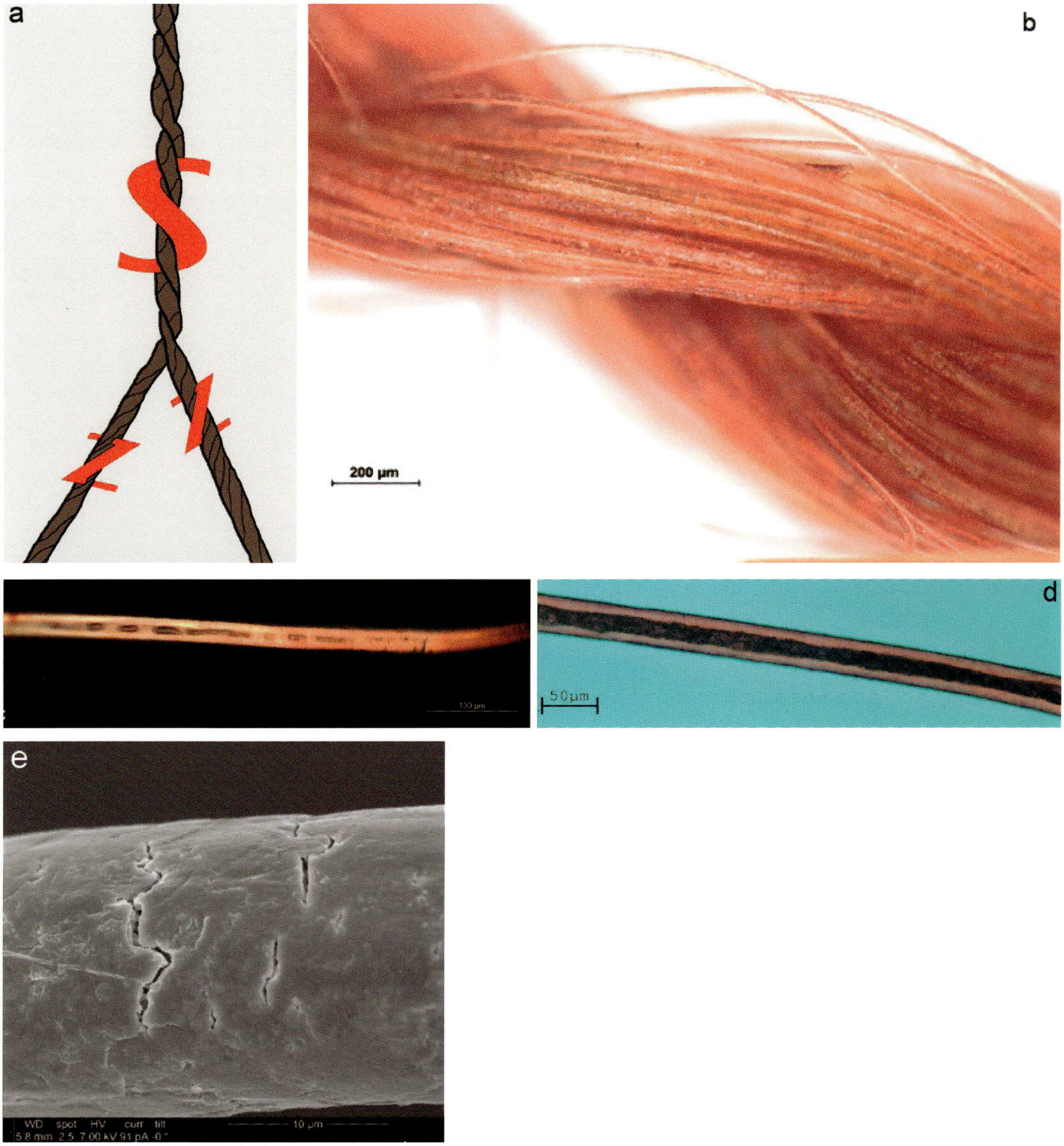

Figure 3 Images of archaeological fibres (a) diagram of the ply and twist pattern of the threads; (b) digital micrograph of a thread from sample VM04; (c) micrograph of sample T-130A in dark field showing the broken medulla; (d) micrograph sample T69 in bright field with an example of an unbroken medulla; (e) scanning electron micrograph of sample T-130A fibre surface showing cracks. Magnifications are indicated by the scales in each micrograph.

The supernatants were injected directly into the HPLC–DAD system.

The HPLC–DAD instrumentation consisted of an 1100 series Agilent Technologies G1322A degasser, G1311A quaternary pump, a Rheodyne 7725i manual injector with a 20 µL loop and a G1315A diode array detector. Dye separation was performed on an Agilent Poroshell 120 SB-C18 reverse-phase column (4.6 × 75 mm, 2.7 µm particle size), which was used with a Poroshell SB-C18 pre-column (4.6 × 5 mm, 2.7 µm particle size). The volume of sample injected was 20 µL. A 20 µL aliquot of methanol was run as a solvent blank before each sample. The blank run was subtracted from that of each sample.

Table 2 Results of analyses.

Sample	Fibre	UV-Vis	HPLC–DAD	SERS
VM01 red	Z spun, 2S ply, camelid	Similar to cochineal	Cochineal	Inconclusive
T-130A	Z spun, 2S ply, camelid	Similar to cochineal	Cochineal	Inconclusive
VM04 red	Z spun, 2S ply, camelid	Similar to cochineal	Cochineal	Inconclusive
061.02.01 red	Z spun, 2S ply, camelid	Similar to cochineal	Cochineal	Inconclusive
061.02.01 pink	Z spun, 2S ply, camelid	Similar to cochineal	Cochineal	Similar to Cochineal
T69	Z spun, 2S ply, camelid	Similar to cochineal	Cochineal	Similar to Cochineal
77.10.02	Z spun, 2S ply, camelid	Similar to cochineal	Cochineal	Not analysed
119.000.029	Z spun, 2S ply, camelid	Similar to cochineal	Cochineal	Not analysed

For the mobile phase, a gradient of 0.1% (v/v) of formic acid in deionised water (A) and 0.1% (v/v) of formic acid in acetonitrile (B) was used. The gradient was 90% A for 3 min, decreasing linearly to 12% in 40 min, and then held for 7 min. The DAD collected data at four wavelengths: 254 nm, 310 nm, 430 nm and 510 nm.

Surface enhanced Raman spectroscopy (SERS)

Individual fibres from each sample were exposed to hydrofluoric acid vapour for 5 min in a closed microchamber.[22] After exposure, a 0.5 μL drop of concentrated silver nanoparticle colloid in water was deposited onto the acidified sample together with 0.5 μL of 0.2M potassium nitrate to induce aggregation of the nanoparticles. The silver nanoparticles were synthesised following the procedure of Lee and Meisel, concentrated 20-fold by centrifugation and resuspension in purified water.[23] SERS analysis was performed with a Chromex Senturion Raman microspectrometer with a 20× objective, using a 785 nm laser diode at 1% power (about 0.70 mW) to excite the Raman spectrum. The laser beam was focused onto the nanoparticle aggregates that formed on the fibres in the drop within a few minutes of deposition of the silver colloid.

Results

The results are summarised in Table 2.

Microscopy

The threads were examined under a digital microscope to determine the nature of their twist and spin. All the Huaca Malena yarn samples had been spun in the Z direction and plied with two threads in the S direction (Fig. 3a and b). Microscopic characterisation of individual fibres showed that all the archaeological samples had a straight, smooth profile, and a central medulla with partially interrupted morphology (Fig. 3c and d). SEM clearly revealed the indistinct scale pattern on the fibre surfaces (Fig. 3e). These attributes and the fibre diameters were all consistent with camelid sources.[24] The range of fibre diameters observed was not conducive to identification of the species of camelid used, although the significantly smaller average fibre diameter for one sample, T130A, suggests that it may have been made of fibres taken from a young animal.[25]

FORS

Features in reflectance spectra of dyed yarns are often indistinct, so the locations of dye absorption maxima cannot be determined accurately. This problem may be alleviated by calculating derivatives of absorption spectra.[26] The wavelengths of minima in the second derivative spectra correspond to maxima in the absorption spectra, which are characteristic of the individual dyes. However, the positions of these maxima may be influenced by mordants and other dyeing aids present in the fibres. Distinctions between dyes with similar spectral features may not be possible, thus making dye identification equivocal.[27] Comparison of unknowns with known reference samples is essential.

Spectral data for the archaeological samples were compared to those for the *Relbunium* and cochineal reference yarns. An example is given in Figure 4 where the second derivative of the Kubelka-Munk transform for sample T69 is shown with those for the references. The shape of the derivative spectrum and the wavelengths of the minima for the Huaca Malena sample (570 and 522 nm) are within experimental reproducibility of those for the cochineal (573 and 519 nm), and not those for the *R. hypocarpium* (546 and 504 nm) dyed yarn. Variations in the spectra in the blue region (below about 480 nm) may be due to differences in the original colours of the fibres themselves, to yellowing of the camelid fibres over time and/or to the presence of substances absorbing in the near UV and violet regions of the spectrum.

All the archaeological samples analysed with FORS presented the same spectral features. These results suggest that cochineal is present in the archaeological samples, but the data are insufficient to assert the absence of other red dyes.

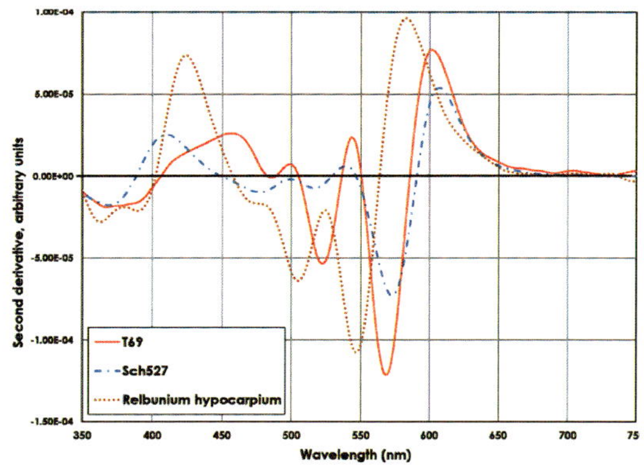

Figure 4 Second derivative of the Kubelka-Munk transform of the smoothed reflectance spectrum of an archaeological sample (T69), compared to those derivatives for reference yarns dyed with cochineal using an alum mordant (Sch 537) or with *Relbunium hypocarpium* prepared by M. Saltzman (see text for sample details and spectroscopic method).

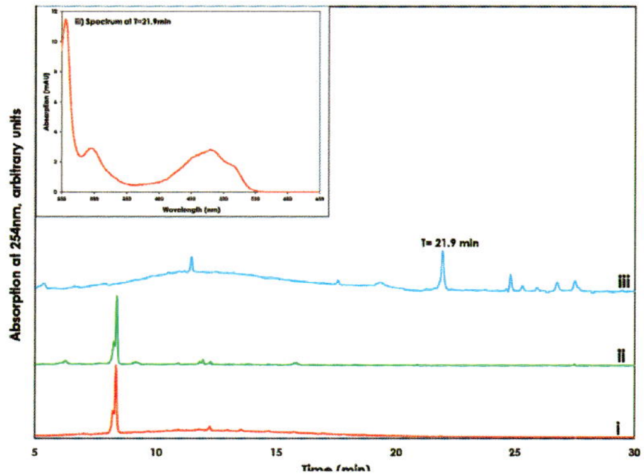

Figure 5 Chromatograms of extracts of two archaeological samples, T69 (i) and VM04 (ii), compared to that of an extract of the *Relbunium hypocarpium* (iii) dyed yarns. The inset shows the absorption spectrum of the coloured peak in the *Relbunium* chromatogram.

HPLC–DAD

Several peaks were detected at 254 nm in the chromatograms obtained by HPLC–DAD analysis of all of the Huaca Malena samples. However, only one major peak in each sample was due to a red-coloured component. When the chromatograms of the archaeological samples were compared to that of an extract from *R. hypocarpium* (Fig. 5), no peak matches were observed. As expected, the chromatogram of the *R. hypocarpium* extract had a strong purpurin peak but no alizarin.

In contrast, comparison of the HPLC–DAD chromatograms of the archaeological sample extracts to that of the cochineal reference solution suggests that all the red-dyed yarns from Huaca Malena contain significant amounts of this dye (Fig. 6a). All the chromatograms presented only the one major, double, peak with a retention time of about 8 min. The UV-Vis spectra of these peaks were all alike: each spectrum had three absorption maxima located at approximately 470 nm, 498 nm and 530 nm (Fig. 6b), corresponding to bands in the absorbance spectrum of carminic acid.[28] Compounds responsible for the minor bands in the chromatograms could not be identified using our instrument and reference library – mass spectroscopy would be required.

SERS

The SERS spectra of archaeological sample T69 and of lighter red fibres of sample 061.02.01

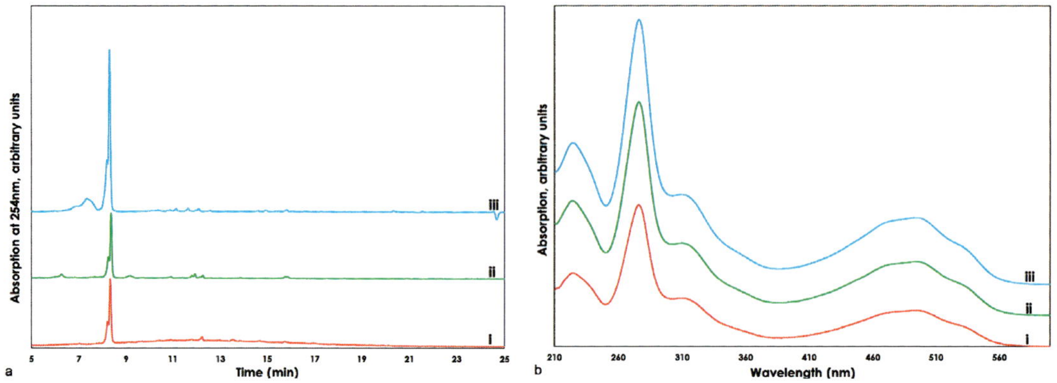

Figure 6 Chromatograms and spectra of extracts of archaeological samples: (a) chromatograms of extracts of T69 (i) and VM04 (ii) compared to that of a cochineal standard solution (iii); (b) absorption spectra of the single coloured peak in each chromatogram.

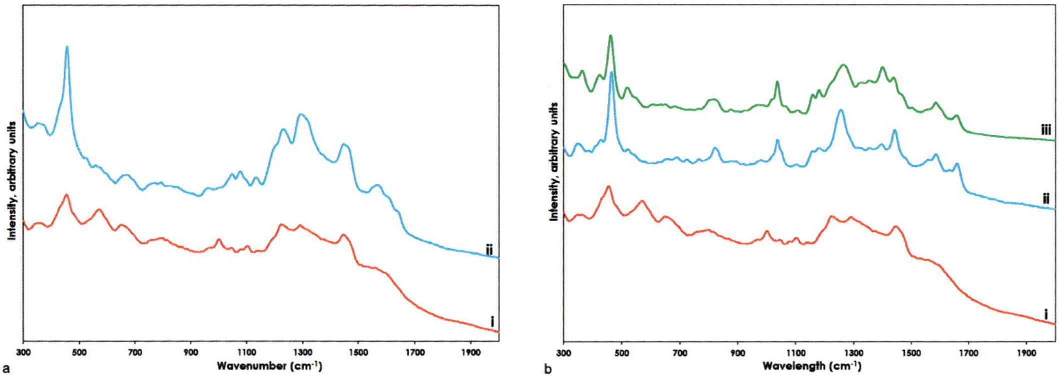

Figure 7 (a) SERS spectra of archaeological sample T69 (i) and Schweppe reference yarn (Sch 527) dyed with cochineal using an alum mordant (ii); (b) SERS spectra of archaeological sample T69 (i) and reference yarns *Relbunium ciliatum* (ii) and *Relbunium hypocarpium* (iii).

(Fig. 7a) were similar to that of the Schweppe reference yarn dyed with cochineal. The spectra of these two archaeological samples presented four bands at 458 cm^{-1}, 1235 cm^{-1}, 1298 cm^{-1} and 1450 cm^{-1}, similar to the bands characteristic for cochineal, despite not being as distinct as the bands for the reference sample. For comparison, the T69 spectrum is also shown with that for the *Relbunium*-dyed reference yarn (Fig. 7b). It can be seen that the archaeological sample has essentially no major Raman bands in common with this reference material. SERS analysis was inconclusive for samples VM01, T-130A, VM04 and darker fibres from 061.02.01.

Conclusion

The finding that all the dyed Huaca Malena yarns were made of camelid fibres is consistent with previous reports of fibres used in patterned pre-Colombian Peruvian textiles from several different cultures.[29] Data obtained by performing HPLC–DAD on extracts from all the archaeological textile samples indicated the use of cochineal as the sole red dye. No bands characteristic for components of other red dye sources, such as *Relbunium* spp., were present in the chromatograms. SERS and FORS corroborated these findings, confirming the HPLC–DAD results. The possibility that additional red dyes were present initially and had degraded completely over time is not likely, but cannot be completely ruled out. It should be noted that the SERS procedure was able to provide useful data in only two of the seven archaeological samples.

Although the Huaca Malena textiles stylistically represent several pre-Colombian cultures, the red yarns were all coloured using cochineal and not *Relbunium* as the dyestuff. This result is consistent with the Wari custom of using this red dye on camelid fibres, and with evidence of Wari dominance of central coast cultures during the latter part of the Middle Horizon period.[30] Plant sources other than *Relbunium* spp. were also used as red dyestuffs by pre-Columbian cultures in Peru.[31] However, our data did not suggest the presence of significant amounts of red dyes other than carminic acid.

Editor's note and authors' addresses

*Author for correspondence: lmaccarelli@lacma.org. Paper received by 13 March 2018; revised version received 16 September 2018. This paper was presented at *Dyes in History and Archaeology* 35, Pisa, Italy, 6–8 October 2016.

[1] Conservation Center, Los Angeles County Museum of Art (LACMA), 5905 Wilshire Blvd, Los Angeles, CA 90036, USA
[2] Western Archeological and Conservation Center, National Park Service Regional Office, Tucson, AZ, USA
[3] Department of Materials Science and Engineering, University of California Los Angeles (UCLA), Los Angeles, CA 90095-1595, USA

Acknowledgements

The authors thank Dr R. Ángeles Falcón for providing the yarn samples from the Huaca Malena textiles to EB, and Dr Charlotte Eng, Conservation Scientist at LACMA for helpful discussion of the manuscript. LM thanks the Andrew W. Mellon Foundation for her Fellowship in Conservation Science. LM and TTS are grateful to the late Dr Frank D. Preusser for early guidance in the project and to Kaye Spilker and Julia Burtenshaw for loan of reference publications. TTS thanks Art Kaplan and Dr Tom Learner of the Getty Conservation Institute for the sample of the cochineal-dyed woollen yarn from the Helmut Schweppe collection. EB thanks UCLA for support and for allowing her to sample the *Relbunium*-dyed camelid yarns in the Max Saltzman collection. EB also thanks Dr Sergey Prikhodko for assistance with SEM and is very grateful to Dr Diana C. Rambaldi for her instruction and guidance on dye analysis and SERS in particular. Research and analysis at UCLA were supported by the Molecular and Nano Archaeology Laboratory at the Henry Samueli School of Engineering and Applied Sciences.

Notes and references

1. Schreiber, K. (2012) 'The rise of an Andean Empire', in S.E. Bergh, *Wari: Lords of the Ancient Andes*, Cleveland, Cleveland Museum of Art, pp. 31–47.
2. Giersz, M., Pardo, C., Makowski, K., Prządka-Giersz, P. and Castro de la Mata, P. (2014) *Castillo de Huarmey: el mausoleo imperial Wari*, Lima, Asociación Museo de Arte de Lima.
3. Schreiber 2012 (see note 1 above).
4. Rodriguez, L.C., Méndez, M.S. and Niemeyer, H.M. (2001) 'Direction of dispersion of cochineal (*Dactylopius coccus* Costa) within the Americas', *Antiquity* 75, pp. 73–7.
5. Bergh, S.E. (2012) 'Tapestry-woven tunics', in S.E. Bergh, *Wari: Lords of the Ancient Andes*, Cleveland, Cleveland Museum of Art, pp. 159–91.
6. Boytner, R. (2006) 'Class, control, and power: the anthropology of textile dyes at Pacatnamu', in M. Young-Sánchez and F.W. Simpson (eds), *Andean Textile Traditions: Papers from the 2001 Mayer Center Symposium at the Denver Art Museum*, Denver, Denver Art Museum, pp. 43–74.
7. Stone-Miller, R. (1992) 'III: Creative abstractions: middle horizon textiles in the Museum of Fine Arts Boston', in R. Stone-Miller, *To Weave for the Sun: Ancient Andean Textiles in the Museum of Fine Arts Boston*, Boston, Museum of Fine Arts, pp. 35–50.
8. Pozzi-Escot, D. and Ángeles Falcón, R. (2011) *Entrelazando el pasado: textiles de Huaca Malena*, Lima, Golder Associates Perú-Universidad Nacional Mayor de San Marcos, pp. 43–60; Bergh 2012 (see note 5 above).
9. Novoa, P. (2000) *Arqueología del valle del Asia: Huaca Malena* (Cuadernos de Investigación del Archivo Tello 2), Lima, Museo de Arqueologia y Antropologia, Universidad Nacional Mayor de San Marcos.
10. Ángeles Falcón, R. and Pozzi-Escot, D. (2000) 'Textiles del horizonte medio. Las evidencias de Huaca Malena valle de Asia', *Boletín de arqueología PUCP*, pp. 401–24.
11. Schreiber 2012 (see note 1 above); Novoa 2000 (see note 9 above).
12. Ángeles Falcón, R. and Pozzi-Escot, D. (2004) 'Del Horizonte medio al horizonte tardío en la costa sur central: el caso del valle de Asia', *Bulletin de l'Institut Français d'Études Andines* 33, 3, pp. 861–86.

13. Frame, M. and Ángeles Falcón, R. (2014) 'A female funerary bundle from Huaca Malena', *Ñawpa Pacha – Journal of Andean Archaeology* 34, 1, pp. 27–59; Pozzi-Escot and Ángeles Falcón 2011 (see note 8 above).
14. Wouters, J. and Rosario-Chirinos, N. (1992) 'Dye analysis of pre-Columbian Peruvian textiles with high-performance liquid chromatography and diode-array detection', *Journal of the American Institute for Conservation* 31, 2, pp. 237–55; Núñez-Regueiro, P., Deliens, J., Desrosiers, S., Oger, B. and Nowik, W. (2006) 'Fragment de tunique Wari', *Techné* 23, pp. 103–6; Niemeyer, H.M. and Agüero, C. (2015) 'Dyes used in pre-Hispanic textiles from the Middle and Late Intermediate periods of San Pedro de Atacama (northern Chile): new insights into patterns of exchange and mobility', *Journal of Archaeological Science* 57, pp. 14–23; Bernard, H. and Boytner, R. (2016) 'The development of Andean textile dyeing technology', poster presented at the Society for American Archaeology 2016 Conference, 7 April 2016.
15. Burr, E. (2016) 'Dye analysis of archaeological Peruvian textiles using surface enhanced Raman spectroscopy (SERS)' (MA thesis, University of California Los Angeles).
16. Cardon, D. (2007) *Natural Dyes: Sources, Tradition, Technology and Science*, London, pp. 159–66; Hofenk de Graaff, J.H. (2004) *The Colourful Past: Origins, Chemistry and Identification of Natural Dyestuffs*, with contributions from W.G.Th. Roelofs and M.R. van Bommel, London and Riggisberg, pp. 124–32.
17. Wouters and Rosario-Chirinos 1992 (see note 14 above).
18. Pozzi-Escot and Ángeles Falcón 2011 (see note 8 above); Frame and Ángeles Falcón 2014 (see note 13 above).
19. Burr 2016 (see note 15 above).
20. Burr 2016 (see note 15 above).
21. Wouters, J., Grzywacz, C.M. and Claro, A. (2011) 'A comparative investigation of hydrolysis methods to analyze natural organic dyes by HPLC-DAD: nine methods, twelve biological sources, ten dye classes, dyed yarns, pigments and paints', *Studies in Conservation*, 56, 3, pp. 231–49.
22. Leona, M., Stenger, J. and Ferloni, E. (2006) 'Application of surface-enhanced Raman scattering techniques to the ultrasensitive identification of natural dyes in work of art', *Journal of Raman Spectroscopy* 37, 10, pp. 981–92; Prikhodko, S.V., Rambaldi, D.C., King, A., Burr, E., Muros, V. and Kakoulli, I. (2015) 'New advancements in SERS dye detection using interfaced SEM and Raman spectromicroscopy (μRS)', *Journal of Raman Spectroscopy* 46, 7, pp. 632–5.
23. Lee, P.C. and Meisel, D. (1982) 'Adsorption and surface-enhanced Raman of dyes on silver and gold sols', *Journal of Physical Chemistry* 86, 17, pp. 3391–5; Rambaldi, D.C., Pozzi, F., Shibayama, N., Leona, M. and Preusser, F.D. (2015) 'Surface-enhanced Raman spectroscopy of various madder species on wool fibers: the role of pseudopurpurin in the interpretation of the spectra', *Journal of Raman Spectroscopy* 46, 11, pp. 1073–81.
24. Tridico, S.R. (2009) 'Natural animal textile fibres: structure, characteristics and identification', in M.M. Houck (ed.), *Identification of Textiles Fibers*, Boca Raton, CRC Press, and Cambridge, Woodhead Publishing, pp. 27 – 67.
25. Burr 2016 (see note 15 above).
26. Karadag, R. and Böhmer, H. (2001) 'Dye analyses using derivative UV-visible spectrophotometry and fibre-safe extraction with EDTA', *Dyes in History and Archaeology* 16/17, pp. 106–18.
27. Schaeffer, T. and Rambaldi, D.C. (2011) 'Are transforms of diffuse reflectance spectra suitable for use in dye identification?', paper presented at the 30th Meeting of *Dyes in History and Archaeology*, University of Derby, 11 October 2011; Gulmini, M., Idone, A., Diana, E., Gastaldi, D., Vaudan, D. and Aceto, M. (2013) 'Identification of dyestuffs in historical textiles: strong and weak points of a non-invasive approach', *Dyes and Pigments* 98, pp. 136–45.
28. Favaro, G., Miliani, C., Romani, A. and Vagnini, M. (2002) 'Role of protolytic interactions in photo-aging processes of carminic acid and carminic lake in solution and painted layers', *Journal of the Chemical Society, Perkin Transactions 2* 2, 1, pp. 192–7.
29. Bergh 2012 (see note 5 above); Stone-Miller 1992 (see note 7 above).
30. Rodrigues *et al.* 2001 (see note 4 above); Bergh 2012 (see note 5 above); Stone-Miller 1992 (see note 7 above).
31. Antúnez de Mayolo, K.K. (1989) 'Peruvian natural dyes plants', *Economic Botany* 43, 2, pp. 181–91.

Ancient Dyes with Medical and Pharmacological Properties

M. Julia Martínez García,[1*] M. Luisa Vázquez de Ágredos Pascual[2]

and Rita Pagiotti[3]

Abstract Vegetal dyes have been described in written sources and medical treatises or pharmacopoeias. They were also prescribed in the recipes of Greek alchemical treatises for the dyeing of textiles, glass and other materials. Since antiquity they have formed components of compounds used as pharmacological remedies. Essential in these formulations were botanical products. Classical sources were essential for learning about these ancient plants and connecting them with those used in the Middle Ages and the modern era in various contexts. Joint analysis of these sources and surviving ancient recipes, found in alchemical treatises and compilations from different periods, has enabled us to confirm the use of certain plant dyes for dyeing purposes and as components of alchemical, pharmacological and magical-medicinal formulations since antiquity. These applications, cited in written sources, are corroborated through research and analysis of rare surviving samples such as those preserved in the Spezieria di Santa Maria della Scala in Rome.

Introduction

Since antiquity, many colouring matters have been used as drugs for reasons including the medicinal properties intrinsic to their composition and the magical conceptions or beliefs associated with them. Often these conceptions and beliefs were influenced by the mineral or organic origin of the pigments and dyes. In the worldview of diverse eastern and western cultures, whether the colouring matters were obtained from the inside of a flower, such as saffron (*Crocus sativus* L.), from a leaf, like the indigo obtained from *Indigofera* species (such as *Indigofera suffruticosa* Mill.), from inside a pod, for example the American achiote (*Bixa orellana* L.), or from the bark of a tree, such as the varieties of brazilwood (including *Haematoxylum brasiletto* H. Karst.), specific meanings directly influenced their social uses. These transcended the artistic, the pictorial and the pigmentary to enter the realms of the medicinal and even the ritual. In ancient Greece, the saffron flower was much more than a dye for colouring fabrics of diverse qualities, such as the woollen jackets of the Minoan women – it was also mixed with red ochre, tallow and beeswax to produce cosmetics such as lipstick. A recipe for perfume found in Knossos, for example, reads: 'Crush the saffron in a crucible with myrrh until it is soft and then mix with oil. Strain three times.' The Greeks, who exported saffron to Egypt, used it for perfume, to dye their hair blonde and to refresh bedclothes. It was also mixed into the cones of perfumed ox fat that men and women wore on their heads.[1] The Romans preserved the dyeing and aromatic uses of saffron and also employed it for ritual and medicinal purposes, which had their origins in ancient Greece. According to legend, saffron was associated with the goddess Artemis, since saffron-coloured tunics were worn by maidens at the

Brauronia ceremony, a rite of passage celebrating the transition from childhood to adolescence and the onset of menstruation,[2] the pain of which could be soothed with tisane made with saffron. The Greek physician Metrodora wrote a medical text entitled *On Women's Diseases* (*c.* AD 200–400) in which she recommended using saffron mixed with pine cone, spikenard and dates soaked in wine to cure haemorrhoids.[3]

This close relationship between colouring matter, art, medicine and rituals persisted throughout the Middle Ages and into the modern era. One important contributory factor was the proliferation of monastic pharmacies during the High and Late Middle Ages and between the Renaissance and the European Baroque. It should be noted, however, that from the 15th century onwards these religious establishments had to compete with secular pharmacies. Rome is home to one monastic pharmacy that preserves the memory of these colorant drugs acquired by doctors and artists (mainly painters and dyers) during the 17th and 18th centuries: the Spezieria di Santa Maria della Scala founded by the Order of the Discalced or Barefoot Carmelites.[4] These substances, which typological studies have identified as originating from the 17th century, are preserved inside glass jars in the dispensing room of the pharmacy. In an adjacent room are the wooden boxes in which the Carmelite friars kept the simple products they used to prepare their compound drugs. These drugs have been analysed using optical microscopy (OM), X-ray diffraction (XRD), X-ray fluorescence (XRF), Fourier transform infrared spectroscopy (FTIR) and gas chromatography–mass spectrometry (GC–MS) for two consecutive projects run by the Department of Art History of the University of Valencia:[5] 'Tracing back to antiquity the composition and significance of ancient drugs, pigments and fragrances found in a 17th-century Roman pharmacy: the archaeometric characterisation and historical-cultural study of an overlooked collection' (2014–2016), and 'Antichi minerali nell'arte degli speziali di "De Medicamentaria Officina" di Santa Maria della Scala, Roma. Indagini Chimico-Fisiche e Studio Storico-Culturale' (2017–2018).

In this paper, we investigate certain mythological associations among a selection of colouring matters used by the Carmelite friars in the 17th and 18th centuries to prepare commonly used drugs. These associations helped to sustain the properties and medicinal uses of the drugs for much of the population.

The ancient Spezieria di Santa Maria della Scala, Rome

The abovementioned project of 2017–2018 focused on the physicochemical and historical study of a collection of 231 pharmaceutical drugs and colouring materials preserved at the ancient Spezieria di Santa Maria della Scala in Rome (Fig. 1). The aim was to analyse the association between the chemical, symbolic and therapeutic composition of these materials and their original equivalents from antiquity. Success has resulted largely from the fact that the research teams have optimised highly precise physicochemical protocols based on the combination of techniques capable of diagnosing organic and inorganic substances mixed in ancient drugs and recipes with therapeutic properties.

The combination of magical-ritual and medicinal properties of most of the colouring materials used to prepare ancient pharmaceutical recipes in Sumer, ancient Egypt, Greece, Etruria and Rome – as well as in ancient India, the Far East and pre-Hispanic America – explains how their physicochemical identification enables us to better understand the cultural and functional aspects of these products in the ancient world and their continued value in modern Europe.

As the appointed pharmacy to various popes, Spezieria di Santa Maria della Scala was famous throughout Europe. Royalty and the nobility also sought remedies for their ailments from the pharmacy's laboratories. The Carmelite monks grew their own herbs in the medicinal garden in the adjacent convent. The prestigious pharmacy soon became popular initially in the local parish community and then among the citizens of Rome in general. In many instances, the label inscriptions on the drug containers point to mythological Graeco-Roman beliefs, while at other times they carry a more modern, that is, 'scientific', name. Studying the collection and precisely characterising the samples enables us not only to trace back

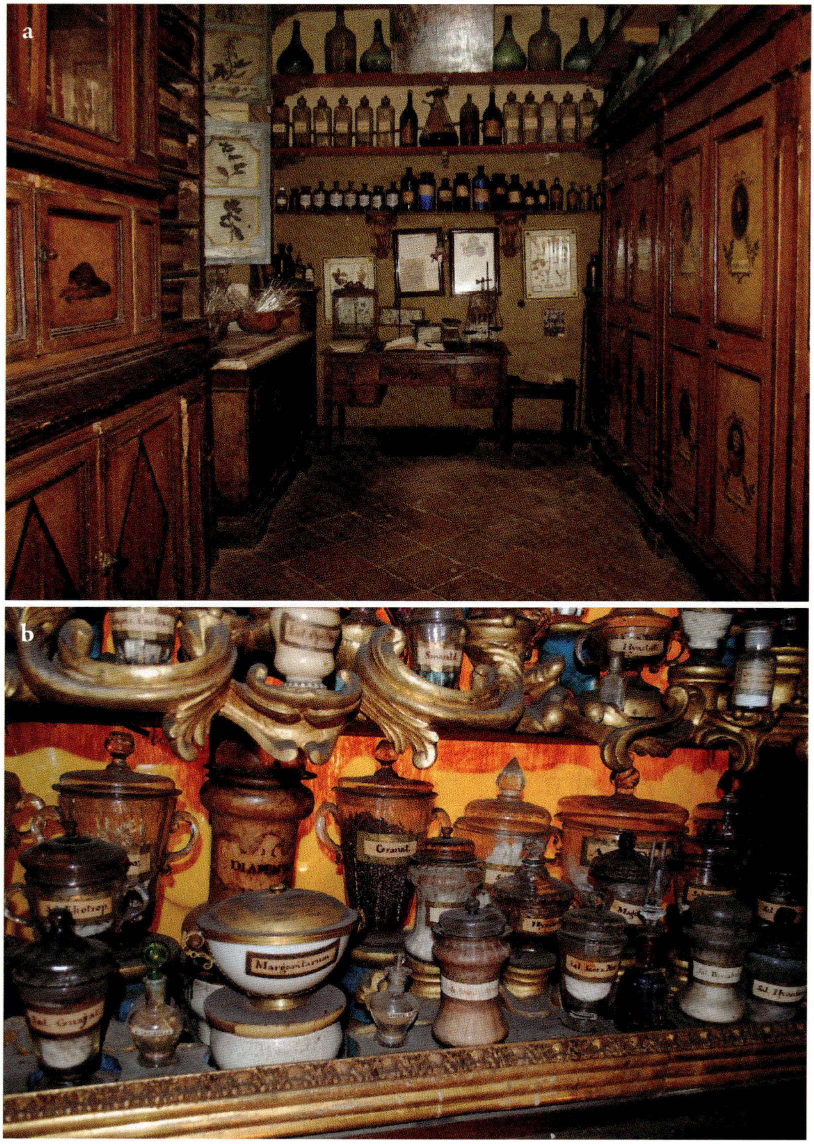

Figure 1 (a) The ancient Spezieria di Santa Maria della Scala, Rome; (b) principal shop vitrine of Santa Maria della Scala. (Photos © Santa Maria della Scala Project)

to antiquity the provenance and intent behind the names, compositions and alleged therapeutic efficacy of the drugs, but also to reconstruct how these drugs were projected into medieval and modern Europe. In other words, to follow the transition from the traditional pre-scientific framework – with numerous mythical and alchemical references – to a more systematic chemical approach.

Study of vegetal dyes located at the Spezieria di Santa Maria della Scala

Studies of vegetal dyes in the historical field are closely linked to the study of ancient botany and we owe the beginning of this science to Theophrastus in about 372–371 BC.[6] The Roman period witnessed the important role played by Dioscorides, a 1st-century AD contemporary of Pliny the Elder, who was both a pharmacist and a botanist. However, it was in the Byzantine era that physicians and herbalists recovered the knowledge of classical antiquity to which the Arabs also contributed their experience. The wealth of Byzantine scholarship during this period is clear.[7]

Fortunately, at the Spezieria di Santa Maria della Scala numerous recipe books (including beautifully decorated herbal manuscripts) and many handwritten letters, formulas, invoices, prescriptions and other documents lay in oblivion in dusty drawers. As with 15th–17th-century manuscripts deposited in other libraries, much of this material contains botanical and pharmacological information.[8] By studying this source of information, we have been able to trace the origins, uses and applications of numerous herbs in pharmacy and medicine. As researchers in dye technology, we were surprised to discover that a dye plant was known as medicinal in antiquity, as well as in medieval and modern times. In this paper, we provide examples of these dye plants, which were used in several fields as textile dyes or for embellishing other artistic supports.

There is evidence that these medicinal plants had many uses in ancient Egypt.[9] We know that perfumes, tinctures and medicinal beverages used sap, fruits, roots, branches, bark, oils and resins of dye species such as saffron, madder, orcaneta or alkanet, lotus, juniper, henbane,[10] safflower and opium as ingredients.[11] These raw materials were already mentioned in the Ebers medical papyrus and in texts from the temples of Edfu and Philae (File).[12]

Plants, shrubs and trees contain several types of compounds. Active ingredients such as vitamins, alkaloids, flavonoids, tannins and antibiotics are responsible for pharmacological actions, dyeing properties and other industrial applications.[13] The chemical fundamentals of the multiple uses of some dyeing plants are due to these same main active ingredients, which must be selectively extracted according to the desired therapeutic or dyeing application. When these plants were used for pharmacological purposes such as healing or poisoning, the flavonoid heterosides, anthocyanins and naphthoquinones in their chemical composition also gave a colour to the products that were prepared using them. Certain extractive pharmaceutical forms were therefore called *tinctures*.

Many plants were also used to prepare magic formulations, charms and poisons. The use of these plants, such as the *Dracaena* (the source of *Sanguis draconis*, dragon's blood), the 'Venus umbilicus' (*Umbilicus pendulinus* DC. = *Umbilicus rupestris* (Salisb.) Dandy, wall pennywort or navelwort in English, *Ombligo de Venus* in Spanish) and the poppy (*Papaver rhoeas* L.), is often steeped in mythology, with qualities generally linked to the most visible properties of plants and associated with heroes or gods who miraculously or accidentally discovered their most remarkable applications. Such legends help us to contextualise the use of these plants and determine the fundamentals underpinning their uses in a given cultural context.

Red dyes

Varieties of red and yellow colours were greatly appreciated in antiquity because of their affinity with blood and gold, respectively, and their corresponding symbolic components. In this study we will therefore focus on two colouring matters: one red, *Sanguis draconis*, or dragon's blood, and one yellow, chamomile flower, both of which were known for their medicinal properties. From the inventory of red-dyeing vegetal materials with pharmacological properties present in the 'simples' cupboards of the pharmacy, such as madder root, alkanet, elderberry and dragon's blood (Fig. 2), we have chosen dragon's blood. Since antiquity, its origin has led to great speculation and it is prescribed in numerous recipes that have survived in technical papyri and other alchemical treatises. It is also found in pharmaceutical formulations of later important pharmacological treatises and modern pharmacopoeias.

Sanguis draconis

Dragon's blood (*Sanguis draconis*) was a popular and highly valuable product in antiquity thanks to its dyeing and medicinal properties. The term '*draco*' has long been used to refer to a venerable tree with a red resin (*S. draconis*) to which wonderful powers were attributed and which played a hugely important role in ancient pharmacopoeias.[14] Dragon's blood was highly appreciated for its rarity but a

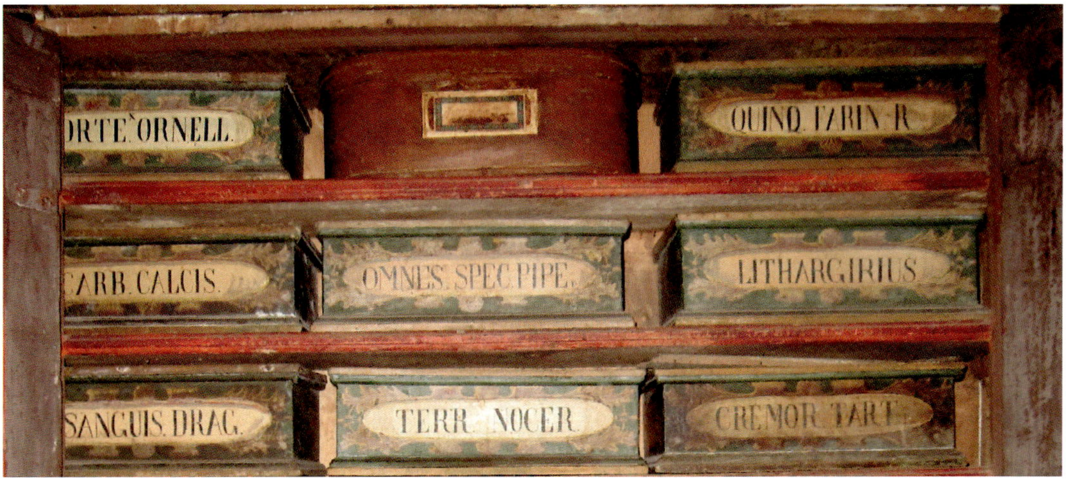

Figure 2 Decorated boxes containing 'simples' stored on shelves at the Spezieria di Santa Maria della Scala. The box for dragon's blood resin (*Sanguis draconis*) is at the bottom left. (Photo © Santa Maria della Scala Project)

lack of knowledge of a specific '*draco*' species plant in antiquity led to much confusion as to its origin. One tradition relates the '*draco*' tree to Geryon's grave and his ancestry. Geryon was a giant who lived on the island of Erytheia in the Gadeiras archipelago, beyond the Pillars of Hercules, who considered himself king of mythical Tartessos.[15] The '*draco*' tree is described as a specimen whose particularity was 'to distil a peculiar blood red juice'. In 1983, a representation of the struggle between Hercules and Geryon was found in Samos on a large metal *lunula* dated from around the 7th century BC. This *lunula* (Fig. 3) depicts several trees with curved trunks and fan-shaped treetops that bear a certain resemblance to twisted palm trees. The trunk of the third tree, which can be seen more clearly, has been identified by some authors as that of a '*draco*' (*Dracaena draco* (L.) L.), which was associated with Geryon.[16]

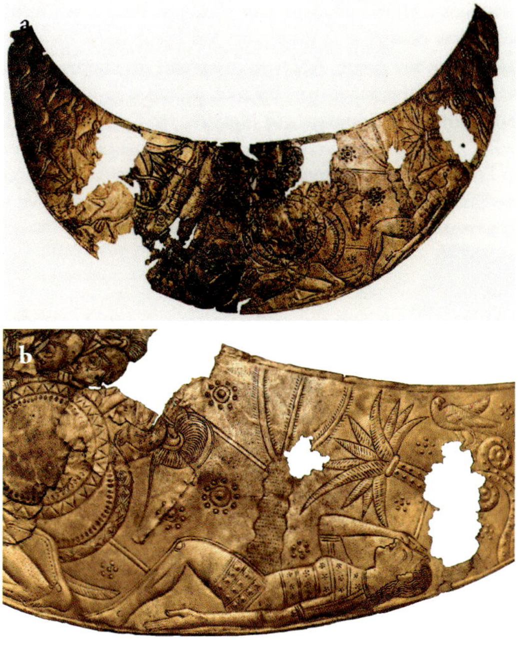

Figure 3 (a) Bronze *lunula* with representation of the struggle between Hercules and Geryon. Bronze, 7th century BC. Vathy Archaeological Museum, Samos, inv. no. A775; (b) detail of the third tree represented in the *lunula*, identified as *Dracaena draco* L. (Photos © Samos Archaeological Museum)

The earliest written reference to the Drago tree seems to come from a text by Strabo (*c.*64 BC–AD 24)[17] in which he records the most accurate description of this plant species from his friend Posidonius (*c.*135–51 BC). The text also states that a tree similar to the Drago, without fruit but with the same curvature of the branches, could be found in Egypt. Philostratus (3rd century AD) reports that among the wonderful and inexplicable things contained in the Temple of Hercules-Gaditanus were trees that they called *Geriones* and that these trees dripped blood. He also briefly described their structure. The legend of dragon's blood and King Geryon indicates that this tree was known by the Phoenicians and other ancient peoples such as the Tartessians.

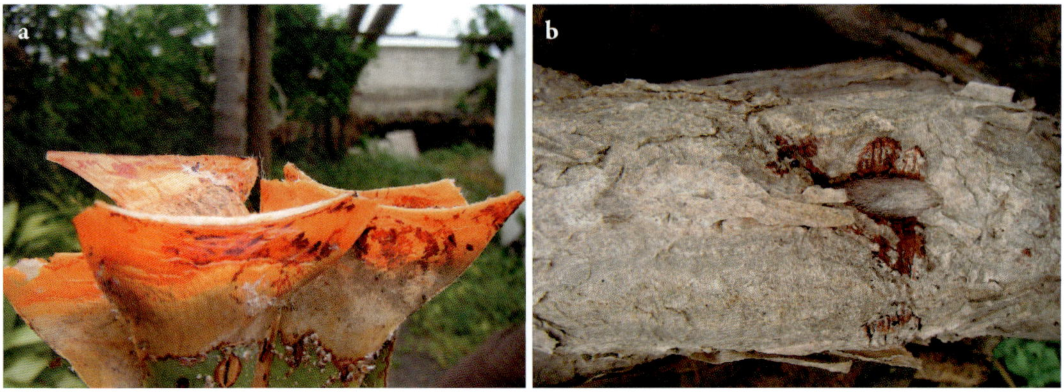

Figure 4 Dragon's blood tree (*Dracaena draco* (L.) L.), La Orotava, Tenerife, Canary Islands: (a) remains of red resin on the leaves and (b) red resin secreted by a cut in the trunk. (Photos © M.J. Martínez García)

Pliny the Elder (1st century AD) informs us of another legend of oriental origin featuring *Sanguis draconis*. This describes the fight to the death between an elephant and a basilisk, a kind of dragon or snake with legs and wings. The basilisk twists around the elephant's body, tearing its skin with its claws and teeth. The loss of blood weakens the robust pachyderm, which crashes to the ground, crushing the strange reptile. In their final throes of agony, the blood of the two beasts combines and when it coagulates it forms an amorphous mass that turns out to have marvellous virtues.[18] Carolus Clusius (Charles de l' Écluse, *c.*1525–1609) had a special interest in exotic plants, including the '*arbor draco*'. His book, written around 1546, contains a strange drawing that represents the fruit of the dragon tree, inside which is a tiny dragon. This drawing illustrates Clusius's criticism of Pliny's legend and reports for the first time the correct nature of dragon's blood, which is simply a tree resin secreted by an incision in the bark or trunk.[19]

Isidore of Seville (*c.* AD 556–638), possibly following Pliny the Elder, claimed that the sap of the *Dracaena draco* of Cadiz turned into gems when it was dried and vitrified.[20] The same reference had been made in medieval times by Al-Himyarí, a 15th-century Arab geographer: both describe the tree as wonderful.[21] 'Drago blood' is traditionally prepared by making a series of incisions in the trunk. The secreted liquid solidifies on contact with the air and acquires a dark red colour, thereby establishing the medieval relationship with gems. After the liquid has dried, it reduces to dust and takes on carmine red hues.[22] This powder was used to prepare varnishes and paints commonly employed to embellish glass with a red marble colour or to redden woods.[23]

Commentators on Dioscorides and modern writers on pharmacognosy have listed various plants with similar names and uses but with different origins and families. These include: *Dracaena cinnabari* Balf.f. from the island of Socotra in the Indian Ocean; *Calamus draco* Willd. which belongs to the palm (Arecaceae) family and originates in India; *Daemonorops draco* (Willd.) Blume, now recognised as a synonym for *Calamus draco* and found in the tropical regions of Southeast Asia; and *Corymbia terminalis* (F.Muell.) K.D.Hill & L.A.S.Johnson, which is native to central Australia.

New World species include *Pterocarpus draco* L. (= *Pterocarpus officinalis* Jacq.), a South American leguminous plant (Fabaceae), and *Croton urucurana* Baill., a native South American tree of the Euphorbiaceae family with a trunk of fine greyish-brown bark that, when wounded, exudes a reddish latex used mainly to dye textiles. Hence the term 'dragon's blood' for these species which dates back to Spanish American colonisation when, during the first expeditions, plant species were identified that secreted a similar resin to the so-called *Sanguis draconis* of the Old World. It was already known in antiquity that this 'dragon's blood' was also obtained from certain trees (Fig. 4) that grew in the so-called 'Fortunate Islands', including *Dracaena draco* (L.) L., whose appearance resembled that of the Socotra island tree.

Humboldt, Berthelot, Schlacht and Mantegazza all talked of an enormous dragon tree in the garden of Sr. Franquí in ancient Taoro (La Orotava) in Tenerife.[24] A Guanche legend asserts that for an incalculable number of centuries this was venerated as a sacred tree, like the olive tree for the Athenians and the Indian banana tree for the Lydians.[25] According to tradition, dragon's blood was among the unguents used by the Guanche during the mummification process, although no archaeological evidence has yet been found to confirm this assertion. José de Viera y Clavijo, the Tenerife historian, states that the dragon tree (*Dracaena draco* L.) is 'a tree famous for its own juice or resin, which sweats from its wounded trunk during the dog days of summer ... since "dragon's blood" is in demand not only from apothecaries but also for its dyes and varnishes'.[26]

Dyeing and colouring applications

The binomial denomination *Sanguis draconis* is used to designate this drug, which was already known by Pliny and Dioscorides. In both Dioscorides' work and the Greek papyri from Roman Egypt, the same denomination was applied to resin from two or three different plants:[27] all are resinous substances with a blood-red hue that are sometimes found in the form of a very fine powder with a crystalline consistency and vitreous appearance. 'Dragon's blood' from *Daemonorops*, a resinous substance with the appearance of fine dust, is extracted from secretions produced by the scale leaves of the fruit of a wide range of palm trees in Southeast Asia. A red resin produced in the trunk of the *Dracaena* is initially liquid and transparent but, on contact with the air, hardens and solidifies to form balls or tears with an intense dark red colour. The characteristic red colour of these natural resins is due to the presence of flavylium compounds such as dracorubin and dracorhodin: more than 20 organic compounds including flavonoids, saponins and chalcones have been identified in its composition, which varies between the different species.[28] A solution of dragon's blood prepared in ethyl alcohol from 100% *Dracaena draco* resin from the Canary Islands gives a more intense dye than a solution from the Southeast Asian material produced in the same proportions.[29]

Periplus maris Erythraei (*The Periplus of the Erythraean Sea*) discusses the organic pigment known as *cinnabari* (*Draconis cinnabari*), which is native to the island of Socotra; dragon's blood from this source is known as *kinnabari* of Dioscorides. Pliny, on the other hand, called it *Sanguis draconis* and it appears with that Latin name in numerous European medieval texts.[30] According to Pliny: 'There is no other colour that paints blood like this.' This pigment was used to dye skins red in antiquity, to colour native shields made with dragon tree bark or to paint the body as, for example, by the native Guanches of the Canary Islands. Domingo Oliva Taroconte stated that 'Dragon's blood was mixed with clay and the juice of other plants, animal fat, etc. to paint the walls of the caves and even the distinctive signs of the clan and the individual.'[31]

Some authors have claimed that dragon's blood was one of the raw materials used to prepare the dye known as 'Getúlica purple', which was famous in antiquity.[32] Dragon trees on the island of Socotra as well as in Sudan and Somalia, areas frequented by ancient Phoenicians, suggest that this was a product that could be included under the term 'gétula' purple, which was prepared on the Mogador islets, the so-called purple islands.[33] According to Brunello, Phoenician dyers in the second millennium BC obtained bright red dragon's blood from Dracaena.[34] The use of these red resins for preparing varnishes to colour metals also dates back to antiquity – even 19th-century craft treatises cite recipes for the preparation of varnishes containing dragon's blood. For example:

> Dragon's blood, sulphur and water, boil, filter; pour this water into a flask with the metal you wish to colour. Boil and distil. The residue has a yellow colour that dyes the metals the colour of gold. You can also operate with equal parts of aloes, saltpetre and copper sulphide.[35]

These recipes have a similar formulation to others described in the Leiden papyrus X for dyeing metal surfaces, for example PLeid. X, no. 51, a recipe for silver browning with sandarac (i.e. 'realgar', an arsenic sulfide), cinnabar, copper sulfates and iron (*misy*). Similarly, in the alchemical treatise *Physica et Mystica* by Pseudo Democritus (also known as Bolos de Mendes, possibly contemporary with

Theophrastus),[36] we find a similar recipe for dyeing superficially with cinnabar, which may be the *kinnabari* of Dioscorides, that is, dragon's blood. Nowadays, the term 'cinnabar' is used to designate mercury(II) sulfide but, for the ancient Greek alchemists, this name had a more complex meaning: it was also used to refer to mercury, realgar, red metallic oxides and the juice of *Dracaena draco*. In his comments on Dioscorides, Andrés Laguna discusses dragon's blood in the chapter on the nature of cinnabar:

> but cinnabar was brought from Africa, and in such a small quantity that it was barely enough to vary the lines of the painting, because it was bought very expensively. Where its colour is deep, some believed that it was Dragon's blood.[37]

These Greek alchemical treatises are associated with a science that is full of mysteries and religious harmony. A later example is the work attributed to the monk Theophilus, dated around AD 1110–1140. His *Diversarium artium schedula*, one of the most important technical treatises on the arts and applied arts of the early Middle Ages, refers to dragon's blood used for producing ink: '*Sanguine draconis* intinge aurum et pone in aereo vase.'[38] However, it also appears in a recipe for dyeing metals gold. In this case it is cited as basilisk powder, following the legend that had survived since the times of Pliny.[39] The components of the recipe are red copper, basilisk powder, human blood and vinegar.[40]

The Lucca Manuscript 490 or Codice Carolinus 490, *Compositiones tingenda musiva*, which is preserved at the Biblioteca Statale in Lucca and dates from around the 8th to the 9th century, also preserves several recipes for dyeing glass and cloth analogous to those of the Greek alchemists.[41] One example is recipe 682D 'Concerning dye of genuine sort'. In this recipe, dragon's blood mixed with vitriol is used to dye green: 'Dye of genuine sort without fire. Grease a fine-spun dyed object with vitriol (?) and smear it with unmixed dragon-wort and it will become like a weaker dyed apple-green-coloured madder.' The recipes gathered under no. XXI (including 682C and 682D) contain one describing the use of dragon's blood for dyeing glass purple (*alithinum*).[42] This is also found in recipes of the 12th-century alchemical compilation known as the *Mappae claviculae*, the most substantial manuscript of which was first published by Sir Thomas Phillipps in 1847. In fact, most of the recipes in Lucca Manuscript 490 are also found in the *Mappae claviculae* such as recipe no. 159: 'Making a purple colour [on glass] without fire: colour thin glass pieces, mix and coat them with dragon's blood, and in this way a reddish colour will result.'[43] This treatise contains recipes for making gold ink from gold powder, for example recipe no. 41, which is very similar to one in the abovementioned treatise by Theophilus. However, the Indian origin of dragon's blood is clearly indicated here: 'Dye gold [powder] with Indian dragon's blood and put in a glass pot. Surround it with coals, and it will immediately liquefy and become fluid enough for you to write with it.'[44] The oldest dragon's blood-producing species, *Daemonorops draco* (Willd.) Blume (= *Calamus draco*, Willd), grows in Sumatra and in the past was accepted as being the only species to produce this substance in India. It was designated by Cordemoy as 'Real dragon's blood from the East Indies', since in the author's opinion it was of better quality than the product obtained from the *Dracaena* species.[45]

Another *Mappae claviculae* recipe for glass dyeing should also be noted. This is recipe no. 69, which gives glass a metallic appearance: 'Giving to glass the nature of a stronger metal.' According to classical literary tradition, this invention dates from the time of Emperor Tiberius. The recipe was subsequently transmitted from Petronius in the 1st century AD to Isidore of Seville and later rediscovered in the Middle Ages.[46] In the Arab world this substance is called *al-ajawayn*, meaning 'blood brothers'. It became well known and was marketed in the medieval Islamic world, including at the caliphate of al-Andalus.

The experience of goldsmiths, sculptors and painters is reflected in important treatises such as that by Cennino Cennini, the earliest extant version of which dates from around 1437.[47] According to Cennini, as González Araña points out, 'dragon's blood, a dye used to dye gold with a reddish colour, was extracted from a seed similar to Avignon grain', which, no doubt, is why the author refers to the Indian product. Resin obtained from specimens of *Daemonorops* genus was sold under the name '*resin draconis*', which may also be the 'Indian dragon's blood' that appears as recipe no. 41 for producing ink with gold dust in the *Mappae claviculae* and is cited by Cennini. In fact, many of the dragon's blood recipes from antiquity until the beginning of the modern age were concerned with the dyeing of metal to obtain a golden colour.[48] Interestingly, dragon's blood was used on wooden statues during the Spanish Baroque period to imitate human blood and was one of the components usually employed to represent Christ's blood on wooden carvings.[49]

The 17th century saw the rediscovery of dragon's blood for use in glass tinting processes. In his *Theatricum Botanicum* of 1640, the English botanist John Parkinson dedicated a chapter to the '*Draco Arbor*' in which he explained the use of dragon's blood by goldsmiths and glassmakers;[50] according to Parkinson, goldsmiths utilised it for glazing purposes. In jewellery, dragon's blood was used to fix a thin sheet of aluminium under precious stones supposedly to add greater brightness. Glassmakers, for their part, applied it hot on stained or similar glass to create a crimson colour.

In Europe, dragon's blood was employed throughout the Middle Ages until its use began to decline in the 14th and 15th centuries. In the 14th century, Catalans established trade on the Atlantic coast of Morocco, for example at Arzila, south of Tangier, where, among other exotic products, they sold a resin that they called *sang de dragó* and which appears in documentation intended for dyeing purposes.[51] According to various Arab writers, this came from the island of Socotra. The East Indies are still important producers of dragon's blood today, exporting it mainly from the port of Singapore.

Since ancient times, dragon's blood has been adulterated to increase profits from commercialisation: Pliny, for example, stated that it was adulterated with goat's blood.[52] In his comments on Dioscorides' book, Mattioli also stated that dragon's blood was compressed into loaves mixed with 'goat's blood, crushed brick and dried rowanberries'.[53]

Pharmacological properties

Due to its homeopathic components, it has traditionally been believed that dragon's blood resin prevents bleeding. While flavonoids afford the resin its colouring properties,[54] other compounds, such as the saponins diosgenin (oestrogenic activity), ruscogenin (anti-inflammatory properties), dracogenin (antibacterial properties) and hecogenin (antileucemic)[55] and chalcones demonstrate pharmacological properties.[56] The medicinal properties of these pharmacological compounds became well known in Europe from the 8th century onwards with the spread of Islam (Fig. 5).

The first remedies prepared with dragon's blood reached Europe around the 9th century AD and were available in the pharmacies of al-Andalus in Muslim Spain. Most doctors there knew this material as *sâyyān*. In the 17th century it became known in France, probably through Arab traders from al-Andalus. *Sanguis draconis*, the Arabic *al-alkhwein dam*, was 'the juice of a plant whose virtues are moderately hot and dry in the second degree', according to the surgeon Henri De Mondeville. Ibn Al-Baytar summarises its main virtues as follows:

> Into the wounds of swords and similar weapon ... shortens haemorrhages ... heals fresh wounds and bleeding ... constipates the vagina ... fortifies teeth ... is useful against the abrasions of the intestines ... is astringent.[57]

Gerardus Johannis Vossius, the 16th-century Dutch humanist, alludes to several beneficial properties for 'dysentery, bleeding and belly flow and, topically, for drying and healing ulcers, though its most common use is to strengthen the gums'.

Examples of modern pharmacopoeias in which dragon's blood is prescribed include John Gerard, *The Herball or Generall Historie of Plantes* (London, 1597), where it is recommended for containing the flow, protecting against dysentery, and consolidating the teeth; John Parkinson, *Theatrum Botanicum: The Theater of Plants* ... (London, 1640), in which it is prescribed for cases of gonorrhoea, watery eyes and small burns; and Johann Schroeder, *Pharmacopoeia medico-chymica sive thesaurus pharmacologicus* (first published in Ulm, 1641, several editions thereafter, including a French edition published in Lyon, 1698) that suggests its use for almost everything. According to a 17th-century Italian manuscript published by Ugo Becciani,

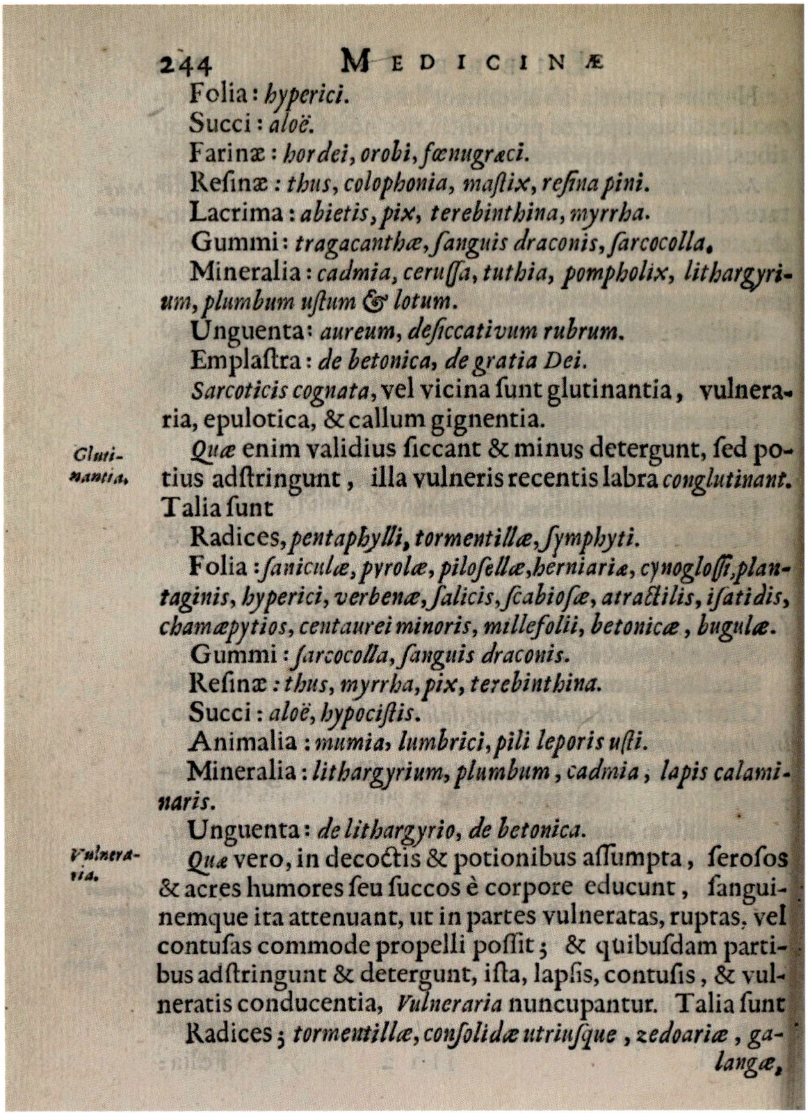
244 MEDICINÆ

Folia: *hyperici.*

Succi: *aloë.*

Farinæ: *hordei, orobi, fœnugræci.*

Reſinæ: *thus, colophonia, maſtix, reſina pini.*

Lacrima: *abietis, pix, terebinthina, myrrha.*

Gummi: *tragacanthæ, ſanguis draconis, ſarcocolla.*

Mineralia: *cadmia, ceruſſa, tuthia, pompholix, lithargyrium, plumbum uſtum & lotum.*

Unguenta: *aureum, deſiccativum rubrum.*

Emplaſtra: *de betonica, de gratia Dei.*

Sarcoticis cognata, vel vicina ſunt glutinantia, vulneraria, epulotica, & callum gignentia.

Glutinantia. *Quæ* enim validius ſiccant & minus detergunt, ſed potius adſtringunt, illa vulneris recentis labra *conglutinant.* Talia ſunt

Radices, *pentaphylli, tormentillæ, ſymphyti.*

Folia: *ſaniculæ, pyrolæ, piloſellæ, herniariæ, cynogloſſi, plantaginis, hyperici, verbenæ, ſalicis, ſcabioſæ, atractilis, iſatidis, chamæpytios, centaurei minoris, millefolii, betonicæ, bugulæ.*

Gummi: *ſarcocolla, ſanguis draconis.*

Reſinæ: *thus, myrrha, pix, terebinthina.*

Succi: *aloë, hypociſtis.*

Animalia: *mumia, lumbrici, pili leporis uſti.*

Mineralia: *lithargyrium, plumbum, cadmia, lapis calaminaris.*

Unguenta: *de lithargyrio, de betonica.*

Vulneraria. *Quæ* vero, in decoctis & potionibus aſſumpta, ſeroſos & acres humores ſeu ſuccos è corpore educunt, ſanguinemque ita attenuant, ut in partes vulneratas, ruptas, vel contuſas commode propelli poſſit; & quibuſdam partibus adſtringunt & detergunt, iſta, lapſis, contuſis, & vulneratis conducentia, *Vulneraria* nuncupantur. Talia ſunt

Radices; *tormentillæ, conſolidæ utriuſque, zedoariæ, galangæ,*

Figure 5 Pharmacological formulation including *Sanguis draconis* (listed as a gum): Henricus Regius (Hendrik de Roy), *Medicina et praxis medica, medicationum exemplis demonstrata*, 3rd edn (Utrecht 1668), p. 244.

one of the most curious virtues it was claimed to have was an ability to restore virginity to young people, 'per la verginità delle ragazze'.[58]

In the Middle Ages, a man could not doubt the word of a saint and even less so when he found his very blood in the apothecary. Dragons lived in unknown, distant places in Africa or India from where mysterious hunters would obtain their blood and reach the apothecary by travelling with Arab caravans. Obviously, the magical legend of these terrible dragons, which were capable of causing all kinds of natural misfortunes, and their medicinal blood endured for a long time and was believed to be both real and fantastic. However, between the 17th and 18th century, Pierre Pomet and his son banished these fantasies from pharmacological treatises.[59]

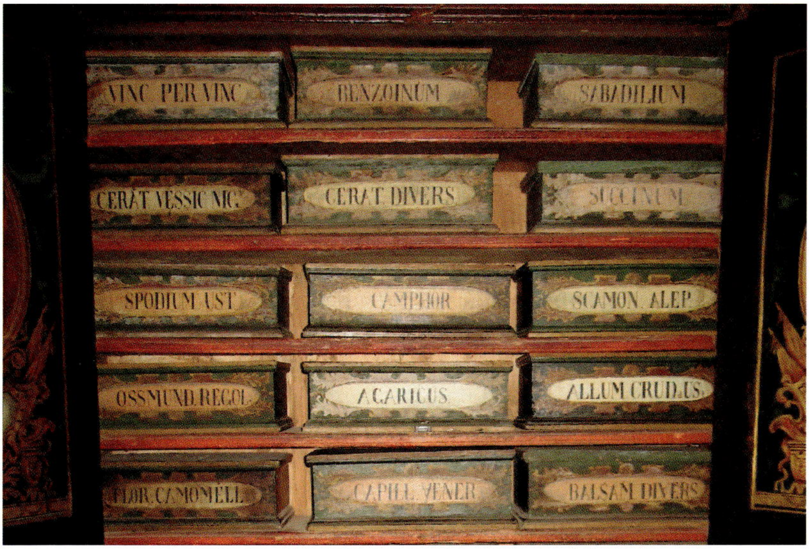

Figure 6 Decorated box of chamomile flowers (*Flor. Camomell*) bottom left, with other boxes of 'simples' at the Spezieria di Santa Maria della Scala, Rome. (Photo © Santa Maria della Scala Project)

Yellow dyes

Vegetal dyes with a beautiful yellow colour and which also have medicinal properties are abundant, as is indicated by the cupboards at Spezieria di Santa Maria della Scala. Examples include saffron, St John's wort, chamomile flower and safflower. Saffron may be considered a paradigm for dyes, rituals and medicines, and much has been written about it, but for this study we focus on the dyeing and medicinal properties of chamomile.

Chamomile flower

There are three known species of chamomile: *Matricaria chamomilla* L., *Anthemis nobilis* L. (a synonym for *Chamaemelum nobile* (L.) All.), and *Anthemis tinctoria* L. (a synonym for *Cota tinctoria* (L.) J.Gay) (Fig. 6). The *Anthemis* genus is characterised by a multiflora, heterogamous flower head (i.e. each plant can have many flower heads and each of these, like other daisy-like species, has two types of flowers): these are the white flowers commonly found in pharmacies and known as the chamomile flower. *A. nobilis* L. (*C. Nobile* (L.) All.) is known as garden chamomile or Roman chamomile. Its white flowers, which are warm and balsamic, have a pleasant aroma and a bitter taste. Wild chamomile flowers should be collected between April and May. Sometimes this variety can be mixed with flowers of other plants of the same genus such as *A. tinctoria* L. or yellow chamomile,[60] which has been used since ancient times to obtain a yellow dye. However, *A. tinctoria* L. belongs to the Asteraceae family and has long been used by the people of Romania for its tinctorial and medicinal properties; pharmacological study of this species is needed due to our limited knowledge. Natural yellow colorants of the flavone group, including apigenin, are present in the chemical composition of *A. chia* L., while luteolin has been identified in *A. tinctoria* L. and other chamomiles such as *A. nobilis* L.[61]

Gardens with a well-developed pharmacopoeia were created close to temples in ancient Egypt as priests understood the properties of medicinal herbs. According to Manniche, yellow chamomile oil was used to embalm the body of Pharaoh Ramesses II. *Anthemis tinctoria* L. was also known to the Egyptians: flower collars made with the yellow chamomile flower were found in the tomb of Tutankhamun and also in the abdominal cavity of Ramesses II. Identified as a dyestuff, its medicinal applications were also known in antiquity. Its name in Coptic is XÀMEMELON.[62]

Theophrastus describes the white chamomile flowers, which he calls *Anthêmon*, and explains that 'there are several varieties of this plant'.[63] Pseudo-Dioscorides distinguished between chamomile and feverfew, which some call *amárakon* and others *anthemís*, *lukánthemon* (of white flower) or *chamaímelon*. The latter term literally means 'apple of the earth' or 'dwarf apple' (from the plant's low-growing habit and scent) and the plant gets its name from the shape and colour of its floral parts. It is interesting to note that chamomile was known throughout the provinces of the Roman Empire, since it implies that its properties were universally known from ancient times: the Gauls, for example, called it *vigneta* and the Dacians, *duodela*. Pseudo-Dioscorides distinguishes another species of chamomile to which he gave the common name 'crazy'.[64] This is simply *Anthemis tinctoria* L., which was known by ancient peoples as *balsamíne*, *Hermoû gónos* (Hermes seed) and *Aphthite gónos* (immortal seed). Africans called it *narôt*.

In all civilisations, nature is closely linked to the world of the gods and can be clearly seen in every reference to medicinal plants in the ancient Mediterranean world. This study aimed to demonstrate the link between mythology and medicine. After several centuries of medicinal practice, therapeutic knowledge became a science. But in ancient times it was believed that the god Hermes named plants and that these were secret names known only to him. The plant known as Hermes seed (*Anthemis tinctoria* L.) is mentioned in the *Papyri Graecae Magicae* (PGM XII, 20, 419). Plants and other objects used in magic formulas were ascribed to images of gods in the encrypted language of magical papyri so that carelessness and mistakes could be avoided in practice and to ensure their effectiveness. In PGM XII, 20, 419, chamomile is therefore called 'hip's blood'.[65] Similarly, certain divinities were closely associated with the life of plants such as *Matricaria chamomilla* L. or a small chamomile known as 'Proserpine grass' since it was one of the plants dedicated to the two infernal goddesses. Feverfew and noble chamomile, on the other hand, were dedicated to the goddess Artemis along with other species such as *Artemisia absinthium* L., for example.[66]

However, it was not only the ancient Greeks who associated these species with the gods: a mythological story about chamomile exists in Norse mythology. Yellow chamomile (*Anthemis tinctoria* L.) is also known as 'Balder's forehead', after Balder, the Norse god and second son of Odin and Frigg. Balder 'the handsome' was venerated as the pure and radiant god of innocence and light who was well versed in science and knew the virtue of flowers. It was called 'Balder's forehead' because of its whiteness (only the pure colour of this flower could be compared to the pure colour of Balder's forehead). The Araucanian Mapuche also have a tradition regarding the origin of the chamomile plant. According to the legend, a young woman put on a beautiful white dress, necklaces, bracelets and a strong yellow headband and waited for nightfall in order to drive away a cougar from the town. In the darkness she was clearly visible in her pale dress (the colour of the chamomile flower) and it was not long before her clothes and necklaces attracted the attention of the huge animal, which began to chase her.[67] The gods rewarded her action by changing her into the chamomile plant. Chamomile is used in infusion form by the herbalist women, or *Machi*, of the Pampa-Patagonian peoples, who are the guardians of all knowledge of health and disease.

In the Middle Ages in the West, chamomile was used to alleviate women's disorders in particular, hence its Latin name *matricaria*, which is based on the belief dating back to the Greek period that plants were associated with virgin goddesses and helped to cure women's diseases. We should also remember that chamomile was one of Artemis's plants and in the medieval period, the Virgin Mary also began to be associated with this plant.[68]

Dyeing applications

We know that yellow chamomile, called chamomile dye or crazy chamomile, has been used since Assyrian times to obtain a strong yellow dye; the Acadians called it *qurban egli* (gift of the field). Depending on the mordant used, this plant was capable of giving orange with a copper mordant, brown with iron, and pure yellow with alum. Ancient botanical treatises such as the book by Theophrastus, *De historia plantarum* (*c.*314 BC), do not specify the application of plants for dyes. And when Dioscorides writes about crazy chamomile, or *káchla*, he mentions a medicinal property (if taken after a bath one regains a good colour, though this may be due to its dyeing properties).[69] Pliny speaks of several varieties of *Anthemis* or chamomile: *Leucanthemis* or white chamomile, which he considers to be the same as *Anthemis*; and *Melanthion*, i.e. *Anthemis tinctoria* or Sibthorpe's dyers' chamomile.[70]

Some recipes from the *Papyrus Graecus Holmiensis* (PHolm.), which dates from around the 3rd and 4th centuries AD, for dyeing wool a yellow colour similar to gold are based on the use of inorganic matter such as litharge, but also contain vegetal materials including 'bull's eye'. PHolm.123 recommends the use of safflower with bull's eye (oxeye daisy) to dye wool a golden colour. The formulation refers to a simple bath in which a mixture of the two flower species is macerated 'to produce a gold colour for a cold bath. Take the flower of the safflower and the bull's eye, chop them together and add water. Plunge the wool, shower with water, take the wool out, dry it and use.'[71] In this recipe, bull's eye may refer to *Anthemis tinctoria* L. Dioscorides called this plant *boúphthalmon* (βούφθαλμον), which refers to its flowers, whose yellow receptacle is eye-shaped.[72] This same Greek term appears in PHolm. 123 and refers to the second dye plant in the recipe.[73]

Authors in the 18th century cite the above plant for dyeing wool yellow. Textile colour was fixed by using sea salt with alum or tin chloride. For purer shades, an animal skin glue mix could be added to the mixture to fix the proteins to the tannin deposited at the bottom of the container; these were eliminated by decantation of the dye liquid. In France, the plant has been cultivated since 1860 for dyeing purposes, and in Germany and Austria it was reintroduced at the beginning of the last century and studied for use in the dyeing of fabrics made of vegetable and animal fibres.[74] Carotenoids are important compounds present in the flowers of *Anthemis tinctoria* L. and the drug is a good source of xanthophylls (the concentration of lutein is 46.8%).[75]

Pharmacological properties

In his medical treatise, Dioscorides describes the medicinal applications of various chamomile species.[76] He informs us that the roots, flowers and plants have a warming property; that in baths they induce childbirth; and that, being emmenagogues, they stimulate menstrual blood flow and could precipitate abortion. They also eliminate jaundice, liver disease and flatulence and help to expel kidney stones. When applied in a cataplasm or poultice, they heal the lacrimal fistula and cure canker sores. Dioscorides cites simple procedures for applying chamomile, such as steam baths, infusions and poultices. The most complicated preparation seems to be an unguent made of flowers dipped in oil to cure fevers. He also described how flowers, leaves and roots should be conserved by mashing them and making them pill-shaped, and stated that the *tinctoria* or yellow-flowered variety is one of the most diuretic, reducing swelling and hardness.[77]

In *De articulis*, the ancient Greek physician Hippocrates (*c*.460–370 BC) prescribes chamomile leaves for fractured head bones, but warns that nothing very cold should be applied. In his book entitled *The Cure of Phrenitics*, Aretaeus, a Greek physician who lived at the time of Nero and Vespasian in the 1st century AD, recommends oil boiled with chamomile (χαμαιμήλῳ) or nard (spikenard) for bathing; and in summer, also 'Hellenic rob' (Ἑλληνικός·), possibly the aromatic calamus or *Acorus calamus* L. (κάλαμος). To cure calculus or kidney stones and kidney ulcers, he suggests filling the bladders of cattle with chamomile oil.[78]

Ancient Arab physicians rediscovered the works of Greek physicians and used them as the basis of their medical teaching. The most famous Arab physician was undoubtedly Avicenna (AD 980–1037), a specialist in pharmacology and medicinal plants, who compiled the most effective remedies of his time. We know that he used chamomile, among other plants. As Dioscorides noted, both *Anthemis nobilis* L. and *Matricaria chamomile* L. or common chamomile have the same medicinal properties, being used as a tonic, carminative and diuretic and for the stomach. Once the plant is prepared, the correct pharmaceutical dosage must be used to administer its active ingredients. The common use of chamomile is as an infusion with half a dozen flower heads per cup.

Because chamomile is composed of more than one active ingredient, it was thought to be extremely effective in the prevention or cure of illness and disease. For this reason, it has been used in numerous pharmaceutical formularies. In 1462, for example, pharmacological preparations with chamomile were plentiful in the medieval pharmacopoeia of a royal apothecary. Included were formulations for unguents, baths, bandages, macerations, embrocations for the head, and so on.[79] These simple pharmaceutical forms had already been described by Dioscorides, but they were still used in medical practice. One example is the following unguent for phlegmon:

> dialtea (2 ounces), wet swab (2 ounces), lily oil (4 ounces), chamomile oil (4 ounces, 113.398 g); total cost = 70 *maravedis*.[80]

The ounce was a unit of mass used in ancient Rome to weigh the goods and other items more accurately, especially if their weight was less than a Roman pound. The *maravedí* was an old Spanish coin in circulation between the 12th and 19th centuries; it was also used as a unit of account. In phytotherapy even today we find formulations containing chamomile in the form of collyrium for curing ophthalmological disease.[81]

As already stated, the therapeutic actions of *Anthemis tinctoria* L. are similar to those of other chamomiles. However, a current study is investigating the traditional uses of this plant in Romania. In animals, the carotenoids present in *A. tinctoria* L. flowers act as antioxidants and tumour-preventing agents, as well as supplying provitamin A. Carotenoids are therefore largely used in the production of food, cosmetics and medicines. As mentioned above, the flower in its role as a drug is a good source of xanthophylls, which have important therapeutic and dyestuff properties.[82]

Conclusion

In ancient Egypt the science that comprised all the secrets of nature was designated 'sacred art', and it was from this sacred art that pharmacy obtained its knowledge. This included the art of preparing the remedies contained in the books that Hermes, considered the inventor of medicine, chemistry and pharmacy, deposited in the temples. At their origin, the arts of healing, pharmacy and the preparation of remedies were influenced by priests. This chemical knowledge also included recipes for metal alloys, pigments and dyes that were analogous to those collected in the Leiden and Stockholm papyri, where the same plants were used as in the preparation of healing and magical remedies.

The healing properties ascribed to animals, minerals and plants were supported by fantastic explanations and legends, as can be seen in relation to the plants studied in this paper. Knowledge was favoured in the Ptolemaic era but, after the fall of the Roman Empire, science declined and plants were only regarded from the point of view of their practical use. It is from the 8th century onwards that the Arabs enriched the medical sciences and shed light on the branches of natural sciences that led to pharmacy.

In the early Middle Ages, the transcription of manuscripts at monasteries led to the emergence of monastic medicine, which collected all current knowledge on medicinal plants. Charlemagne (AD 716–814) ordered convents to cultivate medicinal plants, which led to the development of herbal medicine. In Salerno, between the 10th and 12th centuries, a secular school of medicine emerged that was based on the knowledge of classical authors. The school was made famous by the *Regimen Sanitatis Salernitatum*, which deals with medicinal plants. In the 12th century, numerous recipes, albeit with errors and prejudices, were compiled in the *Compendium Salernitatum*, a manuscript also known as *Collectio Salernitana ossia Documenti inediti e trattati di medicina appartenenti alla scuola medica salernitana.*

In the 13th century the interpretation of medical matters began to change, with philosophers such as Gilbert of England, Pierre de Cresentia, Albert the Great and Aranau de Villanueva advancing the natural sciences and pharmacy. After the mid-16th century, scientific botany began to develop in countries such as France, Germany and Switzerland. Plants were classified in alphabetical order according to their medical and remedial value. Also at this time, the chemist and physician Paracelsus (Theophrastus von Hohenheim, 1493–1541) first related pharmacology to chemistry. As well as plant specimens, on which he was an expert, he used sulfur, lead, antimony, iron and copper in his therapies.

The development of phytotherapy dates from the publication of the first official pharmacopoeia (a list of remedies and officinal preparations) published in Florence in 1498: the *Antidotarium Florentinum*. In France, the reign of Louis XIV saw a chemical and pharmaceutical renaissance. With the support of Jean-Baptiste Colbert, the first minister of state, numerous treatises on chemistry applied to medicine were produced and knowledge of the medicinal properties of plants from antiquity was organised methodologically and rationally, and began to be reinterpreted pharmacologically.[83] Chemical science was becoming specialised. Dyeing and medicinal properties were no longer as closely associated as they had been in ancient treatises. Each plant application was detailed in specific botanical, chemical or pharmacological handbooks depending on their properties.[84]

In the 17th and 18th centuries, the Spezieria di Santa Maria della Scala, known as the papal pharmacy, was famous throughout Europe and used by both royalty and the nobility. On the one hand, in these historic collections, label inscriptions on the medicine containers point in many cases to Graeco-Roman mythological beliefs but on the other, many more modern or 'scientific' names are also found. The study of this collection reveals the transition from a traditional pre-scientific framework, with its numerous mythical and alchemical references, to a more systematic chemical perspective. It follows the historical evolution and rebirth of chemical and pharmacological science, separating the pharmacy of the ritual/magic/religious from the emergence of pharmaceutical chemistry wherever plants are used to treat and cure the sick.

Editor's note and authors' addresses

*Author for correspondence: m.julia.martinez@uv.es. Paper received April 2018; revised version received 8 June 2018. This paper was presented at *Dyes in History and Archaeology* 35, Pisa, Italy, 6–8 October, 2016.

[1] Independent PhD Researcher, external collaborator in the UVEG Project 'Textiles and dyes in the field of ancient clothing: written sources and experimentation' (Dir. Prof. C. Alfaro Giner), Universitat de València, Spain.

[2] Departamento de Historia del Arte, Universitat de València, Spain.

[3] Università degli Studi de Perugia, Italy

Acknowledgements

This research was made possible thanks to the support of the Aboca S.p.A. Società Agricola (Sansepolcro, Italy) through funding for the project: 'Antichi minerali nell'arte degli speziali di "De medicamentaria officina" di Santa Maria della Scala, Roma. Indagini chimico-fisiche e studio storico-culturale' (2017–2018). Similarly, the part of the research on vegetable dyes in antiquity was supported by the ongoing UVEG project, 'Textiles and Dyes in the Field of Ancient Clothing: Written Sources and Experimentation', which began in 2015, and the 'International DressID' project (Culture Program 2007–2012, UE). We are also grateful to Prof. Dr Maarten van Bommel of the University of Amsterdam. Finally we would like to thank the Samos Archaeological Museum, and especially its director, P.J. Chatzidakis, for their kind cooperation.

Notes and references

1. O'Connell, J. (2016) *El libro de las especias: del anís al zumaque*, transl. L. Rins Calahorra, M. Mabres Vicens and J.H. Beutnagel, Barcelona.
2. Vázquez de Ágredos Pascual, M.L. (2011) 'Bajo el signo de Artemisa: iniciación femenina, sexualidad y magia en la Antigua Grecia', in C. Alfaro Giner and M. Albadalejo Vivero (eds), *Las paides y las puellae: aspectos de la infancia femenina en la Antigüedad Clásica*, Valencia, pp. 27–39 (SEMA IX–X).
3. The manuscript, held in the Biblioteca Medicea Laurenziana, Florence, is entitled Περὶ τῶν Γυναικείων παθῶν τῆς μήτρας, Cod. Grec. III (XII century), Plat. LXXV, ff. 4–33; Guerra G. del (1953) *Il Libro di Metrodora; sulla malattie delle donne e il ricettario di cosmetica e terapia*, Greek text with Italian translation, Milan.
4. Vázquez de Ágredos Pascual, M.L., Rojo Iranzo, L., Van-Elslande, E., Walter, P., Pagiotti, R. and Cavallo, G. (2017) 'Science, art and mythological Greco-Roman beliefs in the ancient Pharmacy of Santa Maria della Scala, Rome', in I. Rusu, M. Teodor Nechita, E.N. Drăgoi and N. Apostolescu (eds), *9th European Symposium on Religious Art, Restoration & Conservation: Proceedings Book*, Turin, pp. 116–20.
5. Vázquez de Ágredos Pascual, M.L., Cavallo, G., Pagiotti, R., Rojo Iranzo, L., Souto Martín, M., Walter, P., Van-Elslande, E. and Izzo, F.C. (2018) 'Tradition and renovation in the ancient drugs of the Spezieria di Santa Maria della Scala (Rome):

between scientific knowledge and magical thought', *European Journal of Science and Theology* 14, 2, pp. 3–12.
6. Theophrastus (1988) *Teofrasto, historia de las plantas*, transl. and notes by J.M. Díaz-Regañón, Madrid, pp. 7–11.
7. Thomson, M.H. (1955) *Textes grecs inèdits relatif au plantes*, Paris, pp. 139–68.
8. Bibliothèque nationale de France, Paris, Département des Manuscrits: Ms Paris. gr. 2726, ff. 131r–132r 'Alexandre, Vers à Aglaïs' (on harvesting pepper); Ms Paris. gr. Supplément grec 1301 Année 1840, v. p. 409 (on grafting); Ms. gr., Supplément grec 35, Galen, end 15th century, ff. 185–242 (on the *Mandragora* – mandrake – and other plants); Ms. gr. 825, 16th century, fol. 152 vso, 'Excerpta botanica, iatrica et physica e Praxamo', Julio Africano, Democrito et Florentino (vademecum of plants).
9. Berthelot, M. (1906) 'Archéologie et histoire des sciences. 2. Notices relatives à l'histoire des sciences, IX. Les médecins égyptiens', *Mémoires de l'Académie des sciences de l'Institut de France* series 2, 49, mémoire 2, pp. 1–377: see pp. 232–42.
10. Martínez García. M.J. (2013) 'Cheapening the luxury: some curious recipes with vegetal dyes', in C. Alfaro Giner, J. Ortiz García, and M.J. Martinez García (eds), *Luxury and Dress: Political Power and Appearance in the Roman Empire and its Provinces*, Valencia, pp. 151–69.
11. Stoyanova, M. (2008) 'Written sources on Coptic dyeing methods, Pap. Berol. Copt. 8316: about the meaning of απεϊ from a technical point of view', *Journal of Coptic Studies* 10, pp. 99–112.
12. Berthelot 1906 (see note 9 above).
13. Plant constituents can be grouped as follows: water (60–80% in the leaves and 40–50% in the woody portions); mineral salts (less than 5% of the weight of the dried plant); and organic compounds: carbohydrates, glycosides, lipids, resins and related principles, essential oils, proteins, alkaloids and various other principles (vitamins, organic acids and others). See Stübing Martínez, G. (2013) 'Fitoterapia aplicada', *Anales Reial Acadèmia de Medicina de la Comunitat Valenciana* 14, pp. 1–14, esp. p. 7.
14. Monod, T. (1934) 'Notes canariennes', *La Terre et la vie: revue mensuelle d'histoire naturelle* 4, 8, pp. 466–8.
15. Strabo (1903) [*Geographica*] *The Geography of Strabo*, transl. H.C. Hamilton and W. Falconer, 3 vols, London, George Bell & Sons, III, V, 5. Available at: http://www.perseus.tufts.edu/hopper/text?doc=Perseus:text:1999.01.0239 (accessed 3 February 2021). Other editions are also available online.
16. Brize, P. (1985) 'Samos und Stesichoros. Zu einem früarchaischen Bronzeblech', *Athenische Mitteilungen* 100, p. 5390; Corzo, J.R. (1998) 'El drago de Cádiz en un Bronce Samio del siglo VII a. C.', *Laboratorio de Arte* 11, pp. 27–50, esp. p. 31.
17. Strabo 1903 (see note 15 above), *Geogr.* III, V, 10; Philostratus (1912) *The Life of Appolonius of Tyana*, transl. F.C. Conybeare, 2 vols, London, William Heinemann (Loeb Classical Library), V, 5. Available at: https://www.livius.org/sources/content/philostratus-life-of-apollonius/ (accessed 3 February 2021).
18. Pliny [Gaius Plinius Secundus, called Pliny the Elder] (1855) [*Historia naturalis*] *The Natural History of Pliny*, transl. J. Bostock and H.T. Riley, London, Henry G. Bohn, XXXIII, 38, 39 (subsequent references in the form *NH* [Book], [chapter]). Available at: http://www.perseus.tufts.edu/hopper/text?doc=Perseus%3Atext%3A1999.02.0137%3Abook%3D33%3Achapter%3D38 (accessed 13 February 2021). Other editions, both in Latin and translated, are also available online.
19. Sánchez-Pinto L. and Zárate, R. (2014) 'Sangre de Drago', *Revista Rincones del Atlántico* 6/7. Available at: http://www.rinconesdelatlantico.es/num6/lector.php?id=152 (accessed 11 September 2018, checked 3 February 2021); González Araña, P. (2002) 'Análisis de la resina Sangre de Drago: técnicas y procedimientos artísticos' (PhD thesis, Universidad de la Laguna). Available at: ftp://tesis.bbtk.ull.es/ccssyhum/cs197.pdf (accessed 15 March 2018, checked 3 February 2021).
20. Isidore of Seville (2006) [*Etymologiae*] *The Etymologies of Isidore of Seville*, transl. S.A. Barney, W.J. Lewis, J.A. Beach and O. Berghof, Cambridge, XIV, 6, 7. Based on the Latin edition edited by W.M. Lindsay, Oxford, 1911; Available at: http://penelope.uchicago.edu/Thayer/E/Roman/Texts/Isidore/home.html (accessed 3 February 2021).
21. Lewicki, T. (2012) 'Ibn ʿAbd al-Munʿim al-Ḥimyarī', in *Encyclopaedia of Islam*, 2nd edn, ed. P. Bearman, T. Bianquis, C.E. Bosworth, E. van Donzel and W.P. Heinrichs, Leiden, Brill. Available at: https://referenceworks.brillonline.com/entries/encyclopaedia-of-islam-2/ibn-abd-al-munim-al-himyari-SIM_3030 (accessed 12 September 2018; checked 5 February 2021).
22. Modern techniques of chemical identification (chromatography, Raman spectroscopy, etc.) have confirmed their ancient use in different objects of metal, marble, ceramics, wood etc.: Sánchez-Pinto and Zárate 2014 (see note 19 above).
23. Corzo 1998 (see note 16 above), p. 33.
24. S. Berthelot in his Tenerife description states that the 'dargonero' has been cultivated in the Canary Islands since ancient times, as well as in Madeira and Porto Santo. Likewise, L. de Buch saw it in the wild on the island of Tenerife. They therefore did not believe the theory that this tree was native to the East Indies. The shape of the *Dracaena* is also seen in the Cape of Good Hope, the Bordon island, China and New Zealand: compare Humboldt, A. von (2005) *Viaje a las regiones equinocciales del nuevo mundo: Las Canarias... y otros escritos*, La Laguna, p. 329.

25. Gubernatis, A. de (1879) *La mythologie des plantes, ou, Les légendes du règne végétal*, 2 vols, Paris (1878–9), vol. 2, p. 123.
26. Viera y Clavijo de, J. (1982) *Diccionario de historia natural de las Islas Canarias*, ed. M. Alvar, Las Palmas, p. 16.
27. Berthelot, M. and Ruelle, Ch.- E. (1967) *Collection des anciens alchemistes grecs*, 3 vols, Osnabrück, vol. 2, p. 11.
28. Melo, M.J., Sousa, M., Parola, A.J., Seixas de Melo, J.S., Catarino, F., Marçalo, J. and Pina, F. (2007) 'Identification of 7,4'-dihydroxy-5-methoxyflavylium in "dragon's blood": to be or not to be an anthocyanin', *Chemistry – A European Journal*, 13, 5, pp. 1417–22; Sousa, M.M., Melo, M.J., Parola, A.J., Seixas de Melo, J.S., Catarino, F., Pina, F., Cooke, F.E.M., Simmonds, M.S.J. and Lopes, J.A. (2008) 'Flavylium chromophores as species markers for dragon's blood resins from *Dracaena* and *Daemonorops* trees', *Journal of Chromatography A*, 1209, no. 1–2, pp. 153–61; Baumer, U. and Dietemann, P. (2010) 'Identification and differentiation of dragon's blood in works of art using gas chromatography/mass spectrometry', *Analytical and Bioanalytical Chemistry* 397, pp. 1363–76; Maděra, P., Forrest, A., Hanáček, P., Vahalík, P., Gebauer, R., Plichta, R., Radek, J., Jansen van Rensburg, J., Morris, M., Nadezhdina, N., Vaníčková, L., Jura-Morawiec, J., Wiland-Szymańska, J., Kalivodová, H., Lengálová, K., Rejžek, M. and Habrová, H. (2020) 'What we know and what we do not know about dragon trees?' *Forests* 11(2), 236; https://doi.org/10.3390/f11020236.
29. González Araña, P. 2002 (see note 19 above), pp. 78, 94. To date, most analyses have been carried out on the Canary Islands by scientific teams led by Dr Antonio González and, more recently, by Dr Jaime Bermejo (Institute of Natural Products and Agrobiology, CSIC, Bio-Organic Institute University 'Antonio González', Cancer Research Canary Institute, Natural Sciences Museum of Tenerife).
30. Dioscorides (1998) *De materia medica*, V, 94, 2, in Dioscorides (1998) *Plantas y remedios medicinales (de materia medica)*, introd., transl. and notes by M. García Valdés, 2 vols, Madrid, Gredos, vol. 2, p. 209: 'It is intense in colour, so some thought it was dragon's blood'; Pliny 1855 (see note 18 above), *NH* XXXIII, 38, 39: 'Indeed this last is the only colour that in painting gives a proper representation of blood'. Available at: http://www.perseus.tufts.edu/hopper/text?doc=Perseus%3Atext%3A1999.02.0137%3Abook%3D33%3Achapter%3D38 (accessed 3 February 2021).
31. Oliva Tacoronte, D. (1982) *El drago*, Las Palmas de Gran Canaria, Imprenta Pérez Galdós, p. 15.
32. Martínez García, M.J. (2018) 'Imitaciones, adulteraciones y sucedáneos de la púrpura marina: las noticias de las fuentes y los restos textiles', in M. García Sánchez, and M. Gleba (eds), *Vetux textrinum: Textiles in the Ancient World. Studies in Honour of Carmen Alfaro Giner* (Col-leccio Instrumena 59), Barcelona, pp. 235–53.
33. Tejera Gaspar, A. (2001–2) 'Los dragos de Cádiz y la *falsa púrpura* de los fenicios', *Estudios orientales* 5–6 (Actas II Congreso Internacional del Mundo Púnico, Cartagena, 2000), pp. 369–75.
34. Rinaldi, S. *et al.* (eds) (1986) *La fabbrica dei colori: pigmenti e coloranti nella pittura e nella tintoria*, Rome, p. 111: according to Franco Brunello, Phoenician dyers of the second millennium before Christ were already obtaining the brilliant red colour of the resin exuded by the *dracaena*.
35. Roret, M. (1832) 'Sur l'industrie et les arts et matières', *Collection des Manuels Roret*, vol. 2, Paris, p. 192.
36. Berthelot and Rouelle 1967 (see note 27 above), vol. I, p. 40; see also p. 26 for the nature of sandarac and *misy* in antiquity.
37. Dioscorides, P. (1555) *Pedacio Dioscorides Anazarbeo. Acerca de la materia medicinal y los venenos mortiferos*, transl. with commentary by A. Laguna, Antwerp, V, 68, pp. 539–40. Available at: http://bdh-rd.bne.es/viewer.vm?id=0000037225&page=1 (accessed 3 February 2021).
38. Theophilus (1843) *Théophile, prètre et moine: essai sur divers arts*, ed. and transl. C. de L'Escalopier (Paris), Book I, chap. XXXVII, pp. 60–63, esp. p. 61. Three editions – this one and those by A. Ilg (German) and C.R. Dodwell (English) – are available at http://schedula.uni-koeln.de/ (accessed 5 February 2021). All three have the Latin text as well, but these and other editions vary slightly as to the recipes included and divisions into chapters. The Hawthorn and Smith translation (see note 40 below) has a useful concordance explaining this. Neither this edition nor Dodwell include this particular chapter.
39. Suárez de Ribera, F. (1733) *Pedacio Dioscorides Anazarbeo, annotado por el doctor Andres Laguna*, 2 vols, Madrid, imp. Alonso Balzas, vol. 2, CLXVII, pp. 63–4; C. Cennini (1988) *El libro del arte*, transl. F. Olmeda Latorre, commentary and notes by F. Brunello, Madrid, Akal, CXLIII, pp. 70–71: see Brunello's footnote 1 on dragon's blood.
40. Theophilus 1843 (see note 38 above), Book III, chap. XLVII, pp. 180–81; Theophilus (1979) *On Divers Arts: The Foremost Medieval Treatise on Painting, Glassmaking, and Metalwork*, transl. J.G. Hawthorne and C.S. Smith, New York, Dover (first published Chicago, 1963), Book III, chap. 48, pp. 119–20: 'Spanish gold, which is compounded from red copper, basilisk powder, human blood and vinegar.'
41. Berthelot, M. (1893) *Essai sur la transmission de la Science Antique au Moyen Âge. Doctrines et pratiques chimiques: traditions techniques et traductions arabico-latines*, Paris, p. 11.

42. Burnam, J.M. (1920) *A Classical Technology edited from Codex Lucensis, 490*, Boston, pp. 16, 82; no. 682D, p. 82; no. 682C, p. 82: 'Concerning bloodred dyeing: red blood dyeing. To a pound you put (682C) cinnabar, three ounces and roast for two days': Johnson, R.P. (1939) *Compositiones variae from Codex 490, Biblioteca Capitalare, Lucca, Italy, an introductory study*, Urbana, University of Illinois Press.
43. Phillipps, T. (1847) 'Letter from Sir Thomas Phillipps Bart., F.R.S., F.S.A. ... communicating a transcript of a MS. Treatise on the preparation of pigments ...', *Archaeologia* 32, pp. 183–244: no. 159, p. 218; Smith, C.S. and Hawthorne, J.G. (1974) 'Mappae Clavicula: a little key to the world of medieval techniques', *Transactions of the American Philosophical Society* N.S. 64, 4, no. 159, p. 50.
44. Smith and Hawthorne 1974 (see note 43 above), no. 41, p. 34; no. 69, p. 37: 'Take a sound glass pot, hollow, in the shape of a mortar, and scraping it with an emery stone, scratch all its curved surfaces in a crisscross fashion. Put into a melting furnace some ashes from which glass is to be melted and sprinkle them with raw dragonsblood.'
45. Cordemoy, H.J. (1911) *Les plantes á gommes et á resins*, Paris, p. 309.
46. Petronius Arbiter [C. or T.] (1913) *Satyricon*, transl. M. Heseltine, London, William Heinemann (Loeb Classical Library; published with Seneca's *Apocolocyntosis*), chap. LI. Available at: http://www.perseus.tufts.edu/hopper/text?doc=Perseus%3atext%3a2007.01.0027%3atext%3dSatyricon (accessed 5 February 2021); Pliny 1855 (see note 18 above), *NH*, XXXVI, 66, 67; Dion Casio (Cassius Dio Cocceianus) (1914–27) *Historiae Romanae*, transl. E. Cary, London, William Heinemann (Loeb Classical Library), LVII, 21. Available at: https://penelope.uchicago.edu/Thayer/E/Roman/Texts/Cassius_Dio/57*.html (accessed 5 February 2021); the original Greek text is also available online; Isidore of Seville 2006 (see note 20 above), *Etym.*, XVI, 16; Eraclius (1847) 'De coloribus et artibus romanorum' in M.P. Merrifield, *Original Treatises dating from the XIIth to XVIIIth Centuries on the Arts of Painting*, 2 vols (London), vol. 1, pp. 166–257: Book III, 6, pp. 210–11.
47. This comment is made in the 1956 Spanish edition of Cennini's treatise and is cited by González Araña: Cennini, C. (1956) *Tratado de la pintura: El libro del'arte 1437*, transl. F. Pérez-Dolz, Barcelona, Messeguer, p. 42; González Araña 2002 (see note 19 above), p. 113.
48. Clemente Martínez, C., Salvadó Cabré, N., Butí Papiol, S. and Pradell Cara, T. (2015) 'Estudio de las corladuras sobre oro y plata del retablo barroco de san Rufo de la catedral de Santa María de Tortosa', *Unicum* 14, pp. 181–8, esp. pp. 183–4.
49. Calvo, A. (1997) *Conservación y restauración: materiales, técnicas y procedimientos de la A a la Z*, Madrid, del Serval; Mayer R. (1993) *Materiales y técnicas del arte*, transl. J.M. Ibeas, Madrid, Hermann Blume, pp. 59, 154, 247; González Araña 2002 (see note 19 above), p. 115.
50. Parkinson, J. (1640) *Theatrum Botanicum: The Theater of Plants ...*, London, Tribe 16, chap. 91, pp. 1531–2, *Draco arbor*, the Dragon tree: 'The Goldsmiths and Glasiers use it much in their workes, the one for an enamell, and to set a foile under their pretilous stones, for their greater luster; and the other by fire to strike a crimson colour into glasse, for Windowes or the like.'
51. Dufourcq, C.E. (1966) *L'Espagne catalane et le Maghrib aux XIIIe et XIVe siècles. (De la bataille de las Navas de Tolosa (1212) à l'avénement du sultan mérinide Abou-l-Hassan (1331)*, Paris, Presses Universitaire de France, p. 543; Chaunu, P. (1982) *La expansión europea (siglos XIIl al XV)*. transl. A.M. Mayench, Barcelona, pp. 65–6: 'mes sanch de drago', *Calamus draco* resin, is listed among the products of the African commerce.
52. Pliny 1855 (see note 18 above), *NH*, XXXIII, 39.
53. Mattioli, A. (1554) *Petri Andreae Matthioli Medici Senensis Commentarii, in Libros sex Pedacii Dioscoridis Anazarbei, de Materia Medica, Adjectis quàm plurimis plantarum & animalium imaginibus, eodem authore, o Commentarii*, Lyon, pp. 1355–7: p. 1356, line 20.
54. González, A.G., León, F., Hernández, J.C., Padrón, J.I., Sánchez-Pinto, L. and Bermejo Barrera, J. (2004) 'Flavans of dragon's blood from *Dracaena draco* and *Dracaena tamaranae*', *Biochemical Systematics and Ecology* 32, pp. 179–84.
55. González, A.G., Freire, R., Garcia-Estrada, M.G., Salazar, J.A. and Suárez, E. (1971) 'Nuevas fuentes naturales de sapogeninas esteroidales: XI Dracogenina, nueva tetrahidroxi-sapogenina espirostánica de Dracaena draco L.,' *Anales de Química* 67, 5, p. 557.
56. González, A.G., León, F., Sánchez-Pinto, L., Padrón, J.I. and Bermejo, J. (2000) 'Phenolic compounds of dragon's blood from *Dracaena draco*', *Journal of Natural Products* 63, 9, pp. 1297–9.
57. Sánchez-Pinto and Zárate 2014 (see note 19 above).
58. Sánchez-Pinto and Zárate 2014 (see note 19 above); Becciani, U. *Un manoscritto pistoiese di 'secreti' del tardo '600: lettura critica e commento*, Pistoia, Il Papyrus, 2000.
59. Puerto Sarmiento, F.J. (2009) *La Triaca Magna*, Madrid, Instituto de España, Real Academia Nacional de Farmacia, pp. 21–2.
60. Hanganu, D., Pintea, A., Marculescu, A., Toma, C. and Mirel, S. (2008) 'Chemical research of carotenoids from *Anthemis tinctoria* L. (Asteraceae)', *Farmacia*, LVI, 3, pp. 344–51.
61. Cardon D. (2007) *Natural Dyes: Sources, Tradition, Technology and Science*, London, p. 679.
62. Manniche, L. (2006) *An Ancient Egyptian Herbal*, 2nd rev. edn, London, pp. 18–20, 81.

63. Theophrastus (1916) [*Historia plantarum*] *Enquiry into Plants and Minor Works on Odours and Weather Signs*, transl. A. Hort, 2 vols, London, William Heinemann (Loeb Classical Library), vol. 2, VII, 14, 2, pp. 136–7. Greek and English parallel text. English text available at https://topostext.org/work/242 (accessed 6 February 2021).
64. Pseudo-Dioscorides (1998) [*De herbis feminis*] III, 136–8, 13 in Dioscorides 1998 (see note 30 above), pp. 297–8.
65. Calvo Martínez, J.L. and Sánchez Romero, M.D. (1987) *Textos de magia en papiros. griegos*, Madrid, pp. 274 ff.: compare *Papyri Graecae Magicae* XII, 20, 419 entitled: 'Interpretations translated from the temples served by the sacred scribes'. The *Papyri Graecae Magicae* (Greek magical papyri) are texts from Graeco-Roman Egypt dating from the 2nd century BC to the 5th century AD, which came to light from the early 19th century onwards; the collection is therefore recent. As well as the published editions it can be searched on the Thesaurus Linguae Graecae database: http://stephanus.tlg.uci.edu/index.php (accessed 2 February 2021).
66. Fabre, A.J. (2003) 'Mythologie et plantes médicinales de l'Antiquité', *Histoire des Sciences Médicales* 37, 1, pp. 65–87, esp. pp. 71, 77, 80.
67. Albornoz, A., Farías, V., Montero, G. and Negri, A. (2004) 'Introducción a la complejidad herbolaria de la medicina tradicional mapuche Pampa Patagonia Argentina: análisis multidisciplinario'. (Formerly available at http://www3.unisi.it/cisai/Albetal.doc, or http://www.unisi.it/ricerca/centri/cisai/Albetal.doc (University of Siena Research centres); unavailable 6 February 2021.)
68. Becker, U. (2008) *Enciclopedia de los símbolos*, Barcelona, p. 267.
69. Dioscorides 1555 (see note 37 above), III, 139, p. 356.
70. Pliny 1855 (see note 18 above) *NH*, XXII, 26.
71. Halleux, R. (1981) *Les Alchimistes grecs: tome 1 Papyrus de Leyde, Papyrus de Stockholm – Recettes*, Paris, PHolm. 123, p. 143; *Papyrus Graecus Holmiensis* (PHolm.) is also known as the Stockholm Papyrus: Jensen, W.B. (ed.) (2008) *The Leyden and Stockholm Papyri: Greco-Egyptian Chemical Documents From the Early 4th Century AD*, transl. E.R. Caley, Cincinnati, University of Cincinnati (Oesper Collections in the History of Chemistry, no. 118 (PHolm 118), p. 77: no. 118 in this edition corresponds to no. 123 in Halleux's edition.
72. Dioscorides 1555 (see note 69 above).
73. Halleux 1981 (see note 71 above); Jensen 2008 (see note 71 above).
74. García, M. and Bernard, A.F. (2006) *Plantes colorantes, teintures végétales*, Aix-en-Provence, pp. 32 ff.
75. Hanganu *et al.* 2008 (see note 60 above).
76. Dioscorides 1555 (see note 69 above)
77. Dioscorides 1998 (see note 30 above), vol. 1, Books I–III, p. 468, note 181.
78. Hippocrates (1868) [*De articulis*] *The Genuine Works of Hippocrates*, transl. C.D. Adams, New York, Dover, *On the Articulations*, part 67. Available at: http://www.perseus.tufts.edu/hopper/text?doc=Perseus%3Atext%3A1999.01.0248%3Atext%3DArt.%3Asection%3D67; also https://daedalus.umkc.edu/hippocrates/Adams/page.263.a.php (accessed 7 February 2021),
79. Aretaeus (1972) [*De curatione acutorum morborum libri duo*] *The Extant Works of Aretaeus, the Cappadocian*, ed. and transl. F. Adams (London, 1856), repub. Boston, Milford House: *On the Therapeutics of Acute Disease*, II, III. Available at: http://www.perseus.tufts.edu/hopper/text?doc=Perseus%3atext%3a1999.01.0254%3atext%3dCA (accessed 7 February 2021).
80. Valverde, J.L. and Anibal González, C. (1979) 'Un boticario real de tradición medieval', *Cuaderno de Estudios Medievales* (Departamento Medieval y Ciencias y Técnicas Historiográficas, Universidad de Granada), IV–V, pp. 237–55; compare 'Colirio de manzanilla y hammamelis: Agua de Hammamelis; Agua de Manzanilla amarga; Acido Bórico; Cloruro potásico; Cloruro sódico; Glicerina bidestilada; Cloruro benzalconio; Agua bidestilada csp 100 ml'.
81. Font i Quer, P. (2008) *Plantas medicinales: El Dioscórides renovado*, Barcelona, pp. 807–8, 812. (The author's name is more commonly found as Font Quer.)
82. Hanganu *et al.* 2008 (see note 60 above).
83. Gilbert, E. (1886) *La pharmacie à travers les siècles: antiquité, moyen âge, temps modernes*, Toulouse.
84. Martínez García, M.J. (2014) '*Medicamenta terrena*, materias primas tintóreas en las fuentes antiguas y en la tradición técnico-literaria posterior: herbarios, tratados y recetarios', in L. Rodríguez Peinado and A. Cabrera Lafuente (eds), *La investigación textil y los nuevos métodos de estudio*, Madrid, pp. 25–35; Stübing Martínez 2013 (see note 13 above).

Yellow Dyes of Historical Importance. IV. Sourcing and Consumption of Yellow Dyes by the Dyeing Company of Francesco di Giuliano Salviati e Comp., tintori d'Arte Maggiore in Florence (1483–1498)

Dominique Cardon[1*] and Ingrid Houssaye Michienzi[2]

This article is dedicated to the memory of John Edmonds

Abstract This article offers a direct insight into the importance of yellow dyestuffs in the dye company of Francesco di Giuliano Salviati in Florence, specialised in fast mordant dyes (*arte maggiore*) at the end of the 15th century. It is based on evidence provided by the study of three manuscript account books, respectively dated from1483 to 1486, from 1486 to 1491 and from 1490 to 1498, preserved in the Archivio Salviati, at the Scuola Normale Superiore in Pisa (Italy). A patient deciphering of the books provides exceptionally precise information on what dyestuffs were bought and used; in what quantities each year; how and from where they were obtained; and at what prices. Four yellow dye plants figure in the books and are discussed in this article: sawwort, *Serratula tinctoria* L.; weld, *Reseda luteola* L.; young fustic, *Cotinus coggygria* Scop.; and *guilicie*, a still non-identified plant. Reasons for the use of these plants are proposed and the differences in the colours obtained from sawwort or weld are evidenced by our dye experiments and spectrocolorimetric measurements of the resulting yellow and green samples of wool broadcloth. This study of the three Salviati books reveals that the provenances and consumption of yellow dyestuffs can, and should, be used as indicators of the development of textile production and the evolutions of international markets for European broadcloths.

Introduction

The three previous parts of this long-term multidisciplinary study of yellow dyes of historical importance consisted of historical and chemical surveys of yellow dye plants mentioned in the technical and commercial literature from southern Europe.[1] This fourth part offers a direct insight into the importance of yellow dyestuffs in a Florentine dyeing company that specialised in fast mordant dyes (*arte maggiore*) at the end of the 15th century. It is based on evidence provided by the study of three manuscript account books preserved in the Archivio Salviati, at the Scuola Normale Superiore in Pisa (Italy). The Archivio Salviati is one of the most important archives in Europe for economic and social history. It contains thousands of account books and documents written between the 14th and the 17th century concerning textile production, banking and long-distance trade with many places in Europe and beyond. The present study is but a small part of a vast interdisciplinary research programme based on the study and publication of the archive, under the acronym ENPrESa.[2] Our contribution to this joint effort includes a focus on the documents that concern the history of dyeing.

The three account books mentioned above are the only documents relating to the dye company

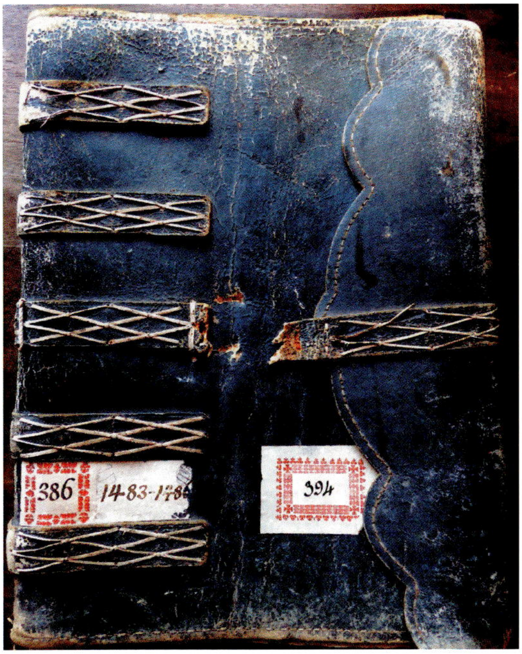

Figure 1 The blue book, labelled A (*Libro azzuro, seg. A*), Reg. 394 in the Salviati Archive, Pisa, registrations from November 1483 to the end of December 1486. (Photo: Dominique Cardon © Archivio Salviati presso Scuola Normale Superiore, *Libro azzuro, segnato A*, Reg. 394)

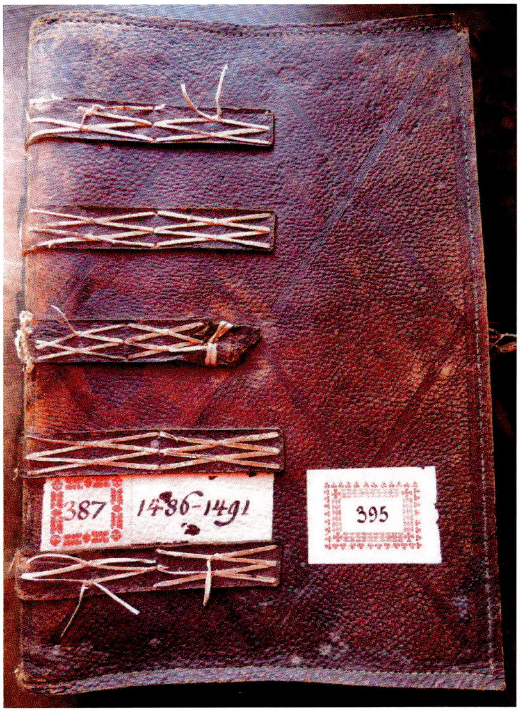

Figure 2 The yellow book, labelled B (*Libro giallo, seg. B*), Reg. 395 in the Salviati Archive, registrations from December 1486 to November 1491. (Photo: Dominique Cardon © Archivio Salviati presso Scuola Normale Superiore, *Libro giallo, segnato B*, Reg. 395)

of Francesco di Giuliano Salviati e Comp., tintori d'Arte Maggiore in Florence that have been preserved. Francesco's father, Giuliano Salviati, together with two of his uncles, Iacopo and Alamanno, was engaged in the production of wool broadcloth of high and medium qualities while Iacopo Salviati was also involved in the production of silk textiles. Francesco, through his dye company, thus completed the activities of an important family business that operated in textile production.

The first two books – Reg. 394, *Libro azzuro, segnato A* (Fig. 1), corresponding to the years 1483 to 1486, and Reg. 395, *Libro giallo, segnato B* (Fig. 2), corresponding to the years 1486 to 1491 – record all the company's commercial dealings organised by products and providers (buying dyes and mordants, paying furnishers, dyers, local custom duties etc.). The third book, Reg. 396, is a *Giornale* (Fig. 3), in which all the daily commercial activities were recorded from 1490 to 1498. No book on the technical aspects of the dyeing activity of the firm having been preserved, all that can be known about the company, therefore, is what dyestuffs were bought and used; in what quantities each year; how and from where they were obtained; and at what prices. All this information allows a better understanding of the nature of the dyeing in which the company specialised.

The lack of any mention of blue dyestuff, either woad or indigo, shows that the company did not include a vat-dyeing division; it obviously specialised in dyeing with fast dyes in boiling baths. On the one hand, the commodities most frequently purchased in large quantities included the typical ingredients of mordanting baths as described in contemporary Italian treatises and regulations on dyeing – alum (*alume*), tartar (*gromma*) and bran (*crusca*) – and on the other, red and yellow dyestuffs. Red dyestuffs recorded include madder (*robbia*), kermes (*grana*), a local *Porphyrophora* sp. 'cochineal' (*chermisi di Marche*), sappanwood (*verzino*), processed orchil and orchil-producing lichens

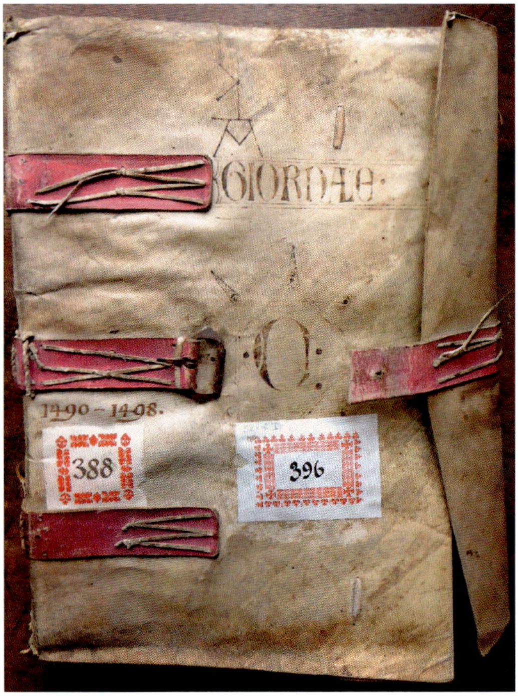

Figure 3 The *Giornale*, diary, Reg. 396 in the Salviati Archive, recording all daily activities from October 1490 to June 1498. (Photo: Dominique Cardon © Archivio Salviati presso Scuola Normale Superiore, *Giornale*, Reg. 396)

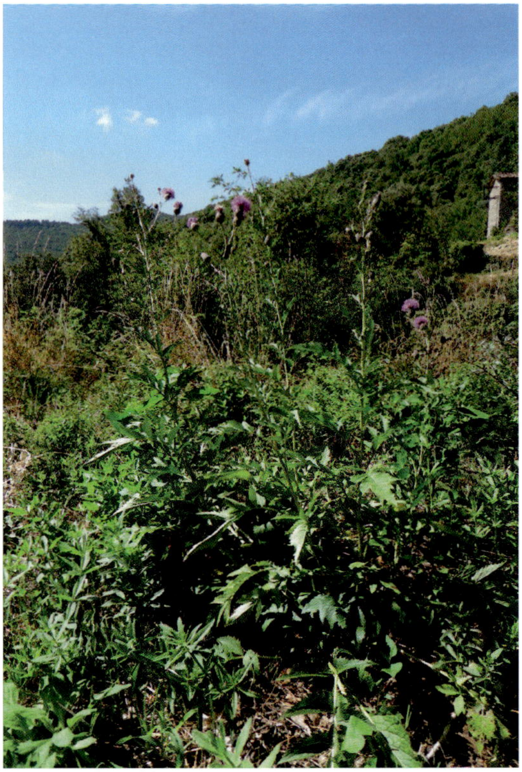

Figure 4 Sawwort plant, *Serratula tinctoria* L. (Photo © Dominique Cardon)

(*oricello* and *raspa*). Four yellow dyestuffs appear regularly in the three books: sawwort, *guilicie*, weld and young fustic.

Sawwort

The first yellow dye plant mentioned in the company's books, *cerretta* (sawwort, *Serratula tinctoria* L.) (Fig. 4) is a tall perennial herb with deciduous upper parts that wither and disappear in the autumn; new leaves do not appear until mid-spring, followed later by floral stems. All of the plant is used for dyeing except the perennial underground part. Sawwort has never been cultivated and had to be collected in its natural habitats during the summer months.[3]

Sawwort is the second dyestuff mentioned in the earliest of the three books, the *Libro azzurro*. It appears initially on folio 4, left and right, immediately following the first double page which is dedicated to the purchases of madder. On the left page (Fig. 5) are recorded all the purchases of sawwort made by the company between November 1483 and December 1486. The first four entries, shown in Figure 6, record purchases of various quantities, all made on the same day – 20 November 1483 – from two types of providers: large amounts were obtained from two distinct middlemen who must have gathered together all the sawwort collected in their area, while smaller quantities were brought by villagers who came to sell plants that they probably harvested either themselves or with a group of relatives.

The following is a transcription of these first records of sawwort purchases and a translation into English. They will serve as representative examples for all entries because all are in the same standardised and much abbreviated format:

> 3518 Cerretta p*er* nostro chonto de' dare, a dì 20 di novenb*re* 1483, F. dieci s. XVI d. I a *oro* l*arghi*, *per* L. 61 s. 12 p*iccioli*, *per* lib. 3 518

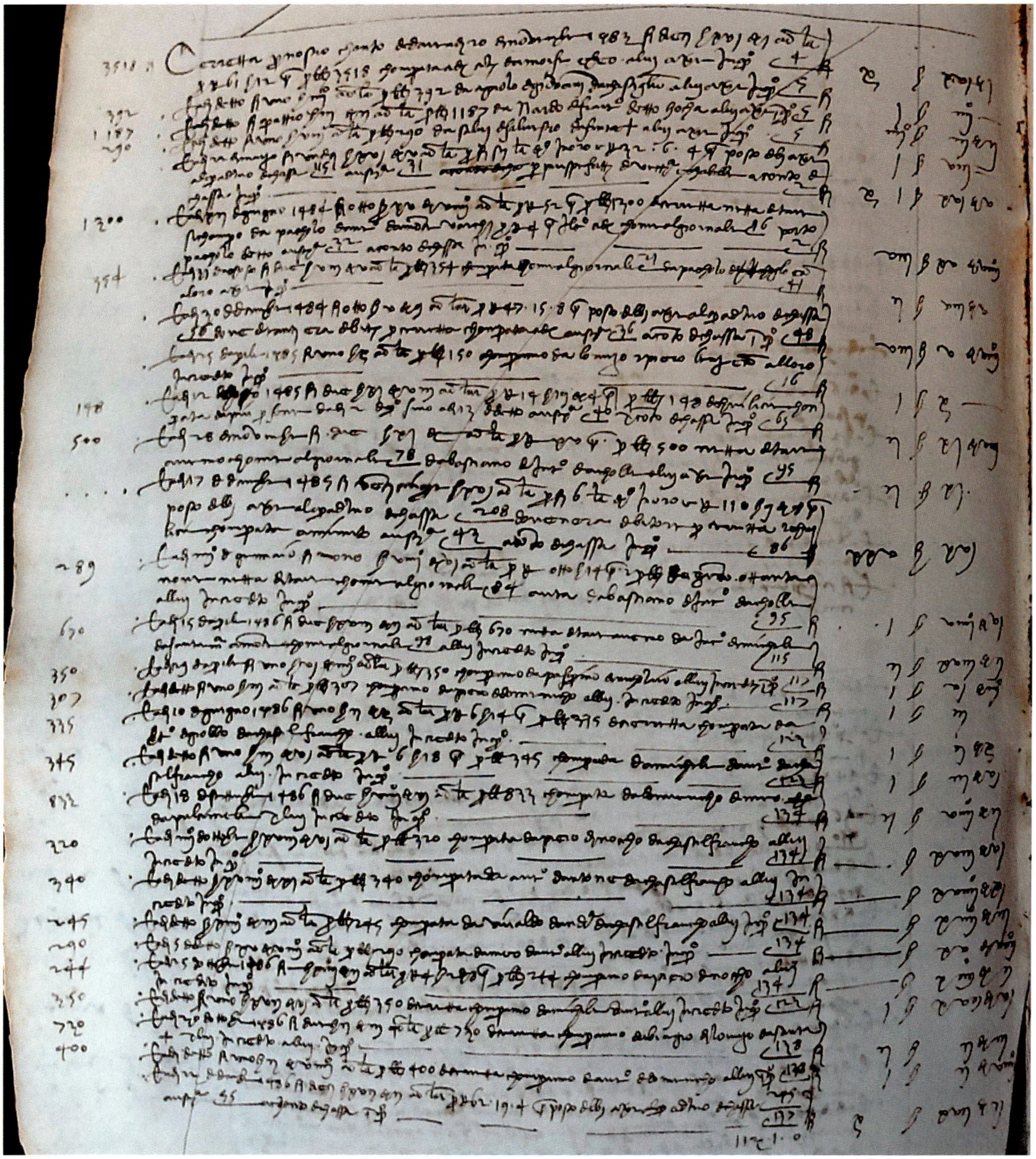

Figure 5 '*Cerretta per nostro conto*', folio 4, left page, of *Libro azzuro, seg. A*, Reg. 394 in the Salviati Archive, recording purchases of sawwort from November 1483 to the end of December 1486. (Photo: Dominique Cardon © Archivio Salviati presso Scuola Normale Superiore, *Libro azzuro, segnato A*, Reg. 394)

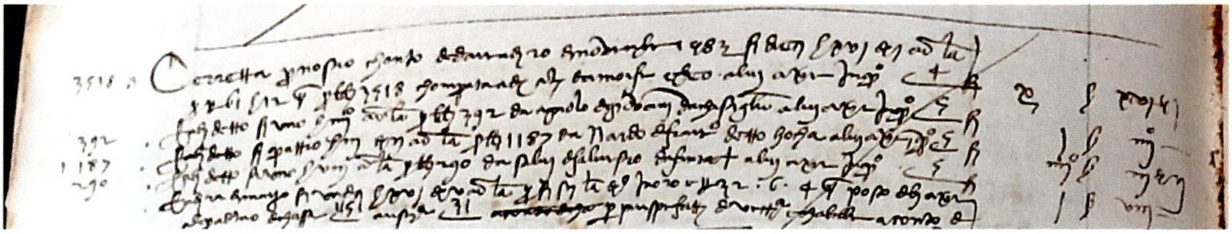

Figure 6 Detail of the first four entries concerning sawwort, on folio 4, *Libro azzuro, seg. A*, Reg. 394 in the Salviati Archive. (Photo: Dominique Cardon © Archivio Salviati presso Scuola Normale Superiore, *Libro azzuro, segnato A*, Reg. 394)

Table 1 Weights of yellow dyes bought by Francesco di Giuliano Salviati e Comp., tintori d'Arte Maggiore from 1483 to 1495, in Florentine pounds (upper line) and converted into kilograms (lower line).

Year	Sawwort + *guilicie*	Weld	Young fustic
1483	5,387 lb[1] 1,826 kg	–	2,820 lb 956 kg
1484	1,654 lb[2] 561 kg	24,232 lb 8,215 kg	4,063 lb 1,377 kg
1485	650 lb[3] 220 kg	40,823 lb 13,859	–
1486	6,048 lb[4] 2,050 kg	17,062 lb 5,784 kg	7,815 lb 2,649 kg
1487	7,911 lb 2,682 kg	33,837 lb 11,471 kg	1,042 lb 353 kg
1488	7,668 lb 2,599 kg	42,519 lb 14,414 kg	2,098 lb 711 kg
1489	5,495 lb 1,863 kg	31,002 lb 10,510 kg	7,116 lb 2,412 kg
1490	2,310 lb 783 kg	37,947 lb 12,864 kg	–
1491	–	46,169 lb 15,651 kg	829 lb 281 kg
1492	–	51,155 lb 17,342 kg	5,103 lb 1,730 kg
1493	–	49,429 lb 16,756 kg	16,550 lb 5,610 kg
1494	–	52,525 lb 17,806 kg	–
1495	–	9,411 lb 3,190 kg	6 *miglia* (6,000 sticks)
Total over 12 years	37,123 lb	436,111 lb	47,436 lb + 6 *miglia*
	12,585 kg	**147,842 kg**	**16,080 kg** + 6,000 sticks

Notes: Kilogram totals may not be exact because of rounding. Purchased on: [1] 20 November; [2] 12 June and 31 August; [3] 25 April and 28 November; [4] between 4 January to 30 October (three intermediary purchases, in April, June, September).

chomp*er*ata a d*enari contan*ti da Moise eb*r*eo; a lui av*ere* in q*uesto* c.4. F. X s. XVI d. I
392 E a dì detto F. uno s. IIII° a o*ro* la*rghi per* lib. 392 da Agnolo di Giovan*ni* da Chastiglio*ne*; a lui av*ere* i*n* q*uesto* c.5. F. I s. IIII°
1187 E a dì detto F. quattro s. III d. II a o*ro* la*rghi per* lib. 1 187 da Nardo di France*sco* detto Hocha; a lui av*ere* i*n* q*uesto* c.5 F. IIII° s. III d. II
290 E a dì detto f. uno s. VIII a o*ro* la*rghi per* lib. 290 da Salvi di Salvestro da Santa †; a lui av*ere* in q*uesto* c.5 F. I s. VIII

[3518[4] Sawwort for our account has to give, on day 20th of November 1483, ten florins *a oro larghi*, 16 *soldi*, 1 *denaro*, for 61 L. 12 s. of *Piccioli*,[5] for 3,518 pounds, bought in cash from Moise, Hebrew; to him to have in this book at folio 4: F. X s. XVI d. I.
392 And on the same day one florin *a oro larghi*, 4 *soldi*, for 392 pounds, from Agnolo di Giovanni, from Castiglione; to him to have in this book at folio 5: F. I s. IIII.
1187 And on the same day four florins, 3 *soldi*, 2 *denari a oro larghi* for 1,187 pounds from Nardo di Francesco, called Hocha; to him to have in this book at folio 5: F. IIII° s. III d. II
290 And on the same day one florin *a oro larghi*, 8 *soldi*, for 290 pounds, from Salvi di Salvestro, from Santa Croce; to him to have in this book at folio 5: F. I s. VIII]

Entries in the three books allow the reader to follow the consumption of sawwort by the company from

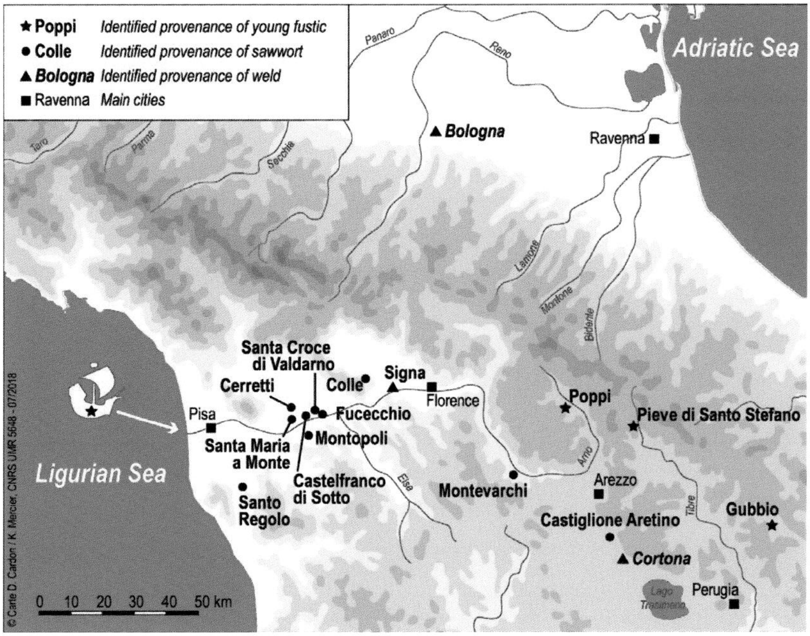

Figure 7 Provenances of sawwort, weld and young fustic in the three account books of the dyeing company Francesco di Giuliano Salviati e Comp., tintori d'Arte Maggiore in Florence (1483–1498). (Map © Dominique Cardon and Karyn Mercier, CIHAM/UMR5648 CNRS)

the second half of November 1483 to mid-September 1490 (Table 1, second column from left).

Over the years, 28 providers from 10 different villages or small towns (Fig. 7) supplied the company with sawwort loads ranging in weight from 180 pounds (61 kg) to 5,000 pounds (1,695 kg).[6] Some suppliers appear only once in the records; others brought two or three loads of sawwort during the same year but do not reappear later; while some were the only providers from their village, coming back for two or three years, bringing their yearly crop of sawwort all at once or in two or three deliveries. None of the villages, however, provided loads of sawwort regularly every year for the whole period. And no mention of sawwort is to be found after September 1490. Explanations for this are proposed below in connection with a comparison between the company's purchases of sawwort and those of weld.

The quantities purchased every year seem impressive at first sight, especially considering that the average weight of a plant of sawwort when dried is about 13.5 g. In 1487, for example, approximately 198,670 plants were collected to obtain the roughly 2,680 kg of dry sawwort bought by Francesco Salviati. The total weight of sawwort purchased between 1483 and 1490 represents approximately 932,200 plants.

In contrast to the details of dyestuff purchases, no information is provided in the Salviati account books on the colours for which sawwort was used or on how much textile would have been dyed with such large quantities of the dyestuff. To answer these questions, quantitative data given above have to be related to other sources, namely two tariffs of the Florentine wool dyers, dated 1428 and 1463. The former sets the price paid to have a fine wool broadcloth dyed green 'with 80 pounds of sawwort' ('*Panno verde fine con 80 lbr. di Cerretta, L. 6 s. 10*'). The 1463 tariff mentions a wider range of colours for which sawwort was used, and that the price had increased, but the quantity of sawwort remains the same: '*Pani Verdi, tanè, gialli, fiordilino e fiordipescho o ranci con 8 lbr. d'allume pro panno e 80 lbr. d'erba ceriecte, L. 7 s. 15*' ('To dye broadcloth green, tan, yellow, flax-flower, peach-flower or orange, with 8 pounds of alum and 80 pounds of sawwort per piece, L. 7 s. 15').[7] If 80 pounds was the standard

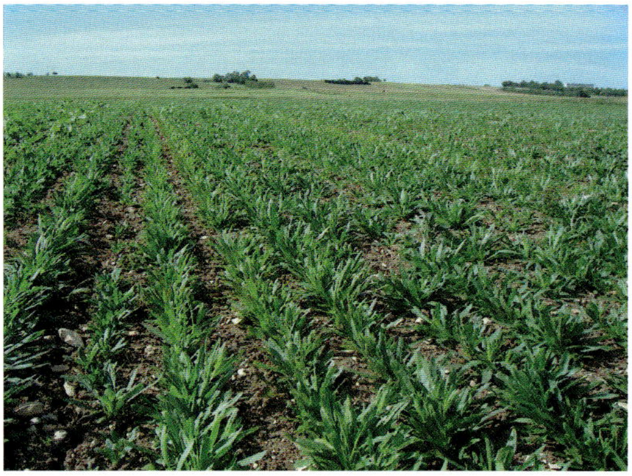

Figure 8 Field of weld (*Reseda luteola* L.) in Charente-Maritime (West France) at the beginning of the 21st century, (Photo © Couleurs de Plantes, Rochefort, published by permission of Anne de la Sayette)

weight of sawwort used to dye a good quality cloth, the total amount purchased by Francesco Salviati's company would have allowed 464 cloth pieces to be dyed one of the colours listed.

Guilicie

One entry, on the same page of the *Libro azzurro* as the records of purchases of sawwort transcribed above, mentions purchases of a different yellow dye plant, *guilicie*, during the first two weeks of August 1485:

> 148 E a dì 12 d'aghosto 1485 F. due s. XI d. VIII a *oro* lar*ghi* *per* L. 14 s. 19 d. 4 p*iccioli* *per* lib. 148 di ghuilicie chon*per*ata da più *per*sone de dì 2 di q*uest*o a dì 13 di detto; a usc*it*a c. 40, a co*n*to di chassa in q*uest*o, c.65. F. II s. XI d. VIII

> [148 And on the 12th of august 1485, two florins *a oro larghi*, 11 *soldi*, 8 *denari*, for 148 pounds of *guilicie* bought from several different persons from the 2nd of this month to the 13th; put on the *uscita* book c. 40, on the cash account in this book c. 65. F. 2 s. 11 d. 8]

The total quantity is not very large, only about 50.2 kg: the price though is much higher – about three times more expensive than the average price of sawwort that year. The inclusion of this record among the purchases of sawwort suggests that *guilicie* was obtained in much the same way as sawwort, i.e. collected in its natural habitats, and had the same uses. The botanical identity of the plant, also mentioned as *erba quilitia* and *guilitia* in other historical sources connected with dyeing, has not been clarified since the first paper on the topic of 'yellow dyes of historical importance' which discussed this issue extensively.[8] The three Salviati books unfortunately do not help in this respect because this unique entry concerning the plant does not mention providers' names or from where they came.

Weld

Gualda or *erba gualda* (weld, *Reseda luteola* L.) (Fig. 8) is a tall biennial herb; to obtain reliable crops, therefore, it needs sowing and cultivation. *Gualda* is frequently mentioned in the three books, from the beginning of 1484 until the end of 1495. The huge quantities purchased show convincingly that it was by far the main yellow dyestuff used in Francesco Salviati's dye-works (Table 1, third column from left).

Weld arrived as whole cartloads of bales of dry plants weighing 207 pounds (70 kg) on average. The company obtained its weld from four places but it actually came from only three centres of production (see Fig. 7). The largest and most regular providers were located in Cortona, which appears to have been a major production centre. From 1485 to 1495, Francesco di Vangelista and Company, the most important of the weld suppliers, sent the dyestuff to Francesco Salviati's company in Florence nearly every month (Fig. 9). The total yearly amounts varied from 2,790 pounds (945 kg) in 1485 to 33,136 pounds (11,230 kg) in 1490. The total quantity of weld purchased during that period from Cortona amounted to 247,519 pounds (83,910 kg). Cortona also appears in the three Salviati books as a principal centre of production for madder, huge quantities of which were bought by the company. Although only 57 out of 225 entries concerning purchases of madder mention its provenance, half the madder bought from that group came from two companies based in Cortona.

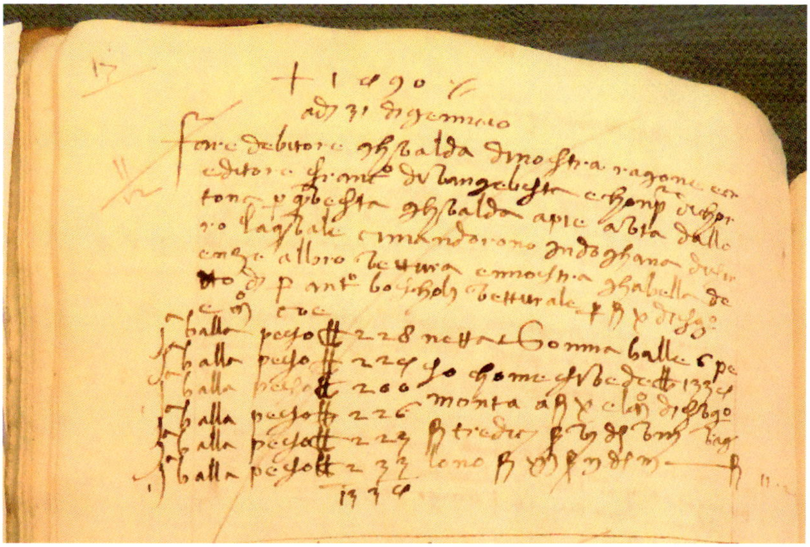

Figure 9 Record of an important purchase of 1,334 pounds (452.893 kg) of weld, *ghualda*, from Cortona, on 31 January 1490. The seller, Francesco di Vangelista, is one of the main providers of dye to Francesco di Giuliano Salviati e Comp., tintori d'Arte Maggiore. The dry weld arrived packed in six bales weighing from 68 to almost 77 kg each. Folio 17v of the *Giornale*, Reg. 396 in the Salviati Archive. (Photo: Dominique Cardon © Archivio Salviati presso Scuola Normale Superiore, *Giornale*, Reg. 396)

Weld was also purchased from suppliers based in Prato and Vernia, near Florence. In Prato, one family provided a total amount of 21,246 pounds (7,200 kg) of weld over the entire period 1484–1495. During the same time, one single provider in Vernia sold a total quantity of 10,657 pounds (3,610 kg) of weld to the company. In both cases, however, records clearly show that this weld, called '*erba bolognese*' or '*gualda bolognese*', actually came from Bologna. Altogether, the purchases of weld from Bologna amounted to 31,903 pounds (10,815 kg). The last provenance mentioned, Signa, also near Florence, appears to have been a centre of production albeit of lesser importance than Cortona and Bologna. Three providers (producers?) from Signa sold a total of 14,125 pounds (4,790 kg) of weld to the company in 1484, and then again in 1487 and 1488. An account book from the dye company of Rafaello di Francesco de' Medici, dated 1502, also mentions weld from Signa ('*erba signesa*'), which confirms that this city was a centre of production.[9]

Weld is not mentioned in the tariffs of the Florentine wool dyers of 1428 and 1463; these sources mention only sawwort as the dyestuff to be used for yellows and greens. But an earlier tariff of July 1375 indicates that to dye *tintillani* broadcloths green, 100 pounds of 'herb' must be used ('*Viridi cum 100 lbr. Erbe*').[10] A parallel to this recipe appeared later in the *Plictho* of Gioanventura Rosetti, which contains a recipe 'To dye yellow' with '*lire 100 de herba gualda*'.[11] Both sawwort and weld are occasionally called 'herb' in the context of yellow and green dyeing: '*erba gualda*' or '*erba cieretta*'.[12] Assuming that the 'herb' mentioned in the tariff of the Florentine wool dyers of 1375 is meant to designate weld and that 100 pounds is adequate to dye a piece of cloth yellow, or green on a blue ground, the total amount purchased by the company would have allowed approximately 4,360 cloth pieces to be dyed yellow or green.

Weld versus sawwort

No weld purchase is recorded for 1483. In the seven-year period from 1484 to 1490, during which both plants were bought and presumably used, the

Table 2 Average prices of yellow dye plants purchased by Francesco di Giuliano Salviati e Comp., tintori d'Arte Maggiore in denari for 100 pounds.

Year	Sawwort	Weld	Young fustic
1483	78	–	235
1484	163	169	167
1485*	200	377	–
1486	83	341	165
1487	97	186	257
1488	141	234	259
1489	131	241	263
1490	100	236	–
1491	–	225	197
1492	–	232	118
1493	–	213	200
1494	–	203	–
1495	–	200	–

* In 1485, *guilicie* was purchased once at a price that would amount to 419 denari for 100 pounds. This is the only time that *guilicie* was recorded as a separate purchase. The other two occurrences mention *guilicie* bought together with sawwort.

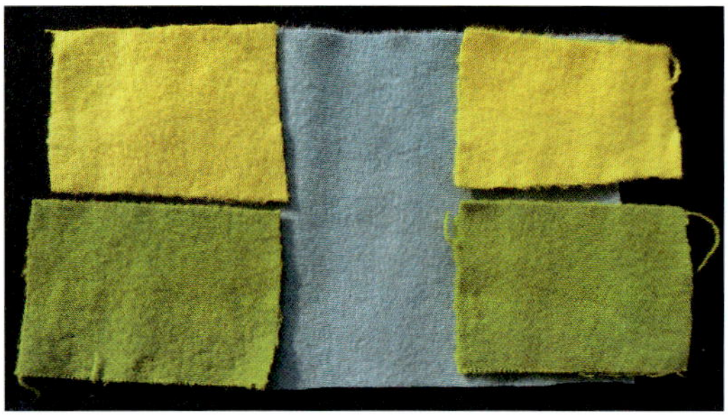

Figure 10 Yellow and green wool broadcloth dyed with weld (*left*) or with sawwort (*right*) placed on a sample of the light blue cloth used for the two greens. Blue wool broadcloth had been dyed in a woad vat by the late John Edmonds, following a Florentine recipe dated 1418. (Photo © Dominique Cardon)

company purchased about 10,760 kg of sawwort and 77,100 kg of weld; sawwort provided 12.2% and weld 87.8% of the total. The average consumption per year can be calculated as approximately 1,535 kg of sawwort and 11,015 kg of weld. Sawwort is no longer mentioned in the books after the end of 1490, but the increase in the purchases of weld the following year, 1491, more than compensated for the lack of sawwort: about 2,785 kg more weld was bought than during the previous year. Why was weld used to a much greater extent than sawwort? The reason is definitely not economic as weld was always more expensive than sawwort and most of the time, significantly so (Table 2).

These data make it obvious that Francesco Salviati had at least one good reason to use sawwort, namely that its cost was much lower than that of weld as a source of yellow and green dyes. Incidentally, this implies that the villagers who collected the plants in their natural habitats (probably mostly women and children) were very poorly paid. The fact that sawwort disappeared from the Salviati books after 1490 is therefore puzzling. Could it indicate that the resource was being exhausted or at least becoming less easily available? Concrete support for this hypothesis – such as company records indicating whether the habitats had been over-exploited and then recovered after the 5–8 years during which sawwort was not listed in the book entries, or if the earlier resources were tapped again at a later date – is absent. A suggestion that this was actually what happened is provided by the account books of a later Florentine dye company, that of Rafaello de' Medici, which record that he bought sawwort on several occasions. Indeed, it was the very first raw material purchased by the firm on 20 May 1502.[13]

Following the same line of thought, if weld, although more expensive, was used in quantities so much larger than sawwort, it was highly likely that it was a more reliable resource, potentially available in any required amount. Recent agronomic research has shown that yields of dry plant per hectare can be in the order of 5,000 kg.[14] Cultivation ensured that weld could be easily collected and delivered in big consignments whenever needed. Another explanation for the disappearance of sawwort from the later Salviati registers, therefore, may simply be that the development of weld cultivation made collecting sawwort from many different locations unnecessarily cumbersome. Ready availability of weld did not make it significantly cheaper, however.

Yellows and greens from sawwort and weld: chromatic characteristics

In addition to the practical issues of sourcing and logistics, other explanations for the choices between the two plants must not be excluded. One possibility to consider is that they may have been related to the chromatic characteristics of the yellow and green colours obtained by dyeing with the two plants.

Table 3 Chromatic characteristics of yellow and green dyes from weld and sawwort.

Colour and dye source	L*	a*	b*	C	h °
Yellow of weld	66.56	0.81	70.09	70.09	89.34
Yellow of sawwort	67.77	1.84	75.47	75.49	88.60
Green of weld on light blue from the woad vat	48.67	-9.0	48.32	49.15	100.55
Green of sawwort on light blue from the woad vat	49.08	-9.18	51.51	52.32	100.11

To check this hypothesis, samples of the same type of wool broadcloth – some undyed and others already dyed a light blue in a woad vat by following a medieval recipe[15] – were washed and mordanted with 16% alum and 1% crude tartar. Half of these samples were dyed with sawwort and the other half with weld in dye baths prepared with weights of dry plant equal to the weights of dry textile to be dyed. The heating of mordant and dye baths and the treatment of the samples of broadcloth in them were conducted under strictly identical conditions for each of the two yellow dyes.

After being washed and allowed to dry in the shade, the samples (Fig. 10) were measured in order to obtain chromaticity coordinates. These data are expressed according to two complementary systems proposed by the Commission Internationale de l'Eclairage (CIE, or International Commission on Illumination): CIE L*a*b*, and CIE L*C*h*. The measurements were performed with a Datacolor microflash colorimeter. Reflectances of the samples were measured between 400 and 710 nm (at 10 nm intervals), using diffuse illumination geometry and a 45°/0° measuring head; specular reflectance was excluded. The colorimetric data corresponding to the CIELab and CIELCh systems were obtained for a CIE 10° standard observer and standard illuminant D65, which corresponds to daylight and is commonly used for textiles. The measured surface area was 59 mm^2 (8.7 mm diameter aperture). Each sample was measured at three different locations. The values presented here (Table 3) are average values calculated from the data obtained by the three measurements.

In the CIELab and CIELCh systems, colour is defined by three co-ordinates:

- Lightness or L* on the axis of greys (from black, corresponding to L* = 0, to white, corresponding to L* = 100)
- On the a* axis, green (negative values of a*) is opposed to red (positive values of a*)
- Along the b* axis, blue (negative values of b*) is opposed to yellow (positive values of b*)

The same system can be expressed in terms of lightness, chromaticity, C* and hue angle, h:

- Lightness is as described above
- C* describes the chromaticity or colourfulness, based on a given level on the axis of greys (C* = 0 = no saturation; C* = 100 = maximum saturation)
- h ° defines hue angle or colour around a circle (0 ° = magenta red; 90 ° = yellow; 180 ° = green; 270 ° = blue; 359 ° = nearly magenta red)

Table 3 shows that the broadcloth dyed yellow with sawwort is slightly lighter and redder than the yellow obtained with weld; it is somewhat more yellow and more saturated. Considering hue, the yellow dyed with weld is an almost perfect yellow, while the yellow of sawwort is very slightly orange. The greens obtained with the two dyestuffs are very similar. The sawwort green is very slightly lighter, greener and more yellow; it is more saturated, but the hues of the two greens are nearly identical. Both greens are very yellow greens.

The somewhat superior yellowness and colourfulness of sawwort, as evidenced by these measurements, could explain why the tariffs of the Florentine wool dyers of 1428 and 1463 mentioned that 80 pounds of sawwort were used to dye a piece of fine wool broadcloth green, while 100 pounds appears to have been the standard weight for dyeing cloth green with weld. These quantities probably corresponded to proportions of dyestuff with respect to the weight of broadcloth to be dyed that were less significant than in our experiments described above, where the yellow dye baths may also have been too concentrated.

Young fustic

Scotano (young fustic, *Cotinus coggygria* L.) is a small deciduous tree or shrub, the heartwood and branches of which are used for dyeing. Like sawwort, it was collected in its natural habitats. It was purchased by Francesco Salviati's company almost every year, most often in smaller quantities than weld. It came in bundles (*fasci*) of branches or thin trunks, counted in *miglia* (1,000 sticks) or measured by weight (Table 1, fourth column from left).

It could be bought in two forms: '*sodo*' (whole sticks) or '*tagliato*' (rasped). Names of various regular employees of the company are mentioned in the books in connection with the rasping (*tagliatura*) of young fustic. Young fustic was purchased both from providers in hilly areas along the Apennine range where the tree is abundant (see Fig. 7), and from other dye companies and Florentine producers of silk textiles (*setaioli*), who used a lot of young fustic for dyeing silk. One big consignment of 73 bundles and one load (*fardello*) of unrasped young fustic with a total weight of 16,562 pounds (5,615 kg), which was registered as 16,550 pounds (5,610 kg) in Francesco Salviati's account book, was obtained in 1493 from the company of Alamanno and Iacopo Salviati through the Salviati branch in Pisa. It was most probably imported by ship from Provence, a region frequently mentioned in medieval merchants' books in connection with young fustic; *scotano provenzalesco* is listed among the commodities sold by the hundred (*centinaio*) in Pisa in Francesco Balducci Pegolotti's *Pratica della mercatura*.[16]

Like sawwort, young fustic was cited as a source of yellow colorants in the tariff of Florentine wool dyers of 1428, but it appeared in connection with a range of colours that are quite different from the yellows and greens of sawwort and weld. Combined with a little madder (10 pounds or 3.39 kg), 60 pounds (20.3 kg) of young fustic were required to dye a piece of fine broadcloth *garofanato* (carnation colour[?]), *tanè* (tan colour) or *rancio* (orange).[17] The Archivio Datini in Prato holds records of actual dyeing processes for broadcloth, some of which are dated 1396 and document the use of young fustic for still more shades: *dorè* (golden), which requires 25.5 pounds (8.64 kg) of young fustic per cloth piece; *pelo di leone* (lion's coat), which requires 27 pounds (9.15 kg) per piece of cloth previously dyed Turkish blue (*turchino*) in the woad vat; and *cotognino* (quince colour), which requires 23 pounds (7.8 kg) per piece of cloth previously dyed *alazato* blue in the woad vat.[18] On the basis of these data, the total amount of young fustic purchased by Francesco Salviati's company could have been used to dye from 790 up to 2,060 cloth pieces, depending on the colours produced.

Conclusion

The significance of the quantitative data obtained through the careful deciphering of the three books from the dye company of Francesco di Giuliano Salviati e Comp., tintori d'Arte Maggiore is easier to understand when the numbers are arranged in the two comparative tables. This provides immediate recognition of the importance of the different sources of yellow dyestuffs used by the company. The quantities recorded in Table 1 are impressive, especially when taking into account the fact that Francesco Salviati's dye company was neither the only nor the most important one in Florence at the time; in the three books, at least 30 other dyers then trading in Florence are listed.

The precise data provided by these three books give an astonishing insight into the huge requirements for dyestuffs that accompanied the massive development of wool broadcloth production in many parts of Europe during the later Middle Ages. The parallel development of the cultivation of the most important dye plants – woad for all shades of blue and blue grounds, madder for all colours needing some red – has so far been studied to a much greater extent than the sourcing of yellow dyes. The three Salviati books reveal that the provenances and consumption of yellow dyestuffs can, and should, also be used as indicators of the development of textile production. In this case, the huge quantities of yellow dyes bought by Francesco Salviati between 1483 and 1495 provide new evidence to confirm the significant rise in Florentine broadcloth production which, Hidetoshi Hoshino has argued, took place in the second half of the 15th century, in contrast to Alfred Doren's vision of the '*decadenza*' of this industry during that period.[19] Hoshino attributed this new development largely to the conquest of the large markets of the Ottoman Empire by Florentine merchants.

The increased need for yellow dyes, necessary to produce greens on various blue grounds, supports this view. The demand was most likely due to the popularity of the various shades of green among the predominantly Muslim populations of the Ottoman Empire, for whom green is the colour preferred by the Prophet Muhammad. An illustration of this has been provided by Ingrid Houssaye Michienzi's recent study of three account books written between 1491 and 1493 in Ottoman Constantinople by Giovanni di Marco Salviati, a distant cousin of Francesco, who imported large quantities of cloth pieces from Florence. Their colour was not always mentioned but, out of 261 pieces for which it was described, 35% were different shades of green, 2 pieces (0.8%) were lemon yellow (*limon*), and 6 more were dyed colours typically obtained with young fustic, *tanè* and tawny (*lionati*).[20] Francesco apparently chose to cater to this target market and specialise in green dyeing; of 721 records of dyeings of wool broadcloth that mention the colour produced, more than 85% were various shades of green.

Editor's note and authors' addresses

*Author for correspondence: cardon.dominique@wanadoo.fr. At the time of writing Ingrid Houssaye Michienzi was employed at Villa I Tatti/The Harvard University Center for Italian Renaissance Studies, Florence, Italy. Paper received 19 August 2018, revised version received by 26 November 2018. This paper was presented at *Dyes in History and Archaeology* 34, Thessaloniki, Greece, 22–23 October, 2015.

[1] CIHAM/UMR 5648 CNRS, ISH, 14 av. Berthelot, 69363 Lyon Cédex 07, France
[2] CNRS, UMR 8167 Orient & Méditerranée, 27 rue Paul Bert, 94200 Ivry-sur-Seine, France

Acknowledgements

We are deeply grateful to Maddalena Taglioli, Curator of the Salviati Archive, and to all the team at the Scuola Normale Superiore in Pisa for their unfailing kindness and helpfulness; to all our colleagues and friends of the ANR ENPrESA programme for very fruitful exchanges; and to Karyn Mercier, CIHAM/UMR 5648, CNRS, for her beautiful map.

Notes and references

1. Cardon, D. (1995) 'Yellow dyes of historical importance: beginnings of a long-term multi-disciplinary study. I: Yellow dye-plants in the technical and commercial literature from southern Europe: Italian, French and Spanish sources of the 13th–18th centuries', *Dyes in History and Archaeology* 13, pp. 59–73; Andary, C., Prunac, S. and Cardon, D. (1996) 'Yellow dyes of historical importance: a multi-disciplinary study. II: Chemical analysis of weld and saw-wort', *Dyes in History and Archaeology* 14, pp. 33–8; Cardon, D. and Andary, C. (2001) 'Yellow dyes of historical importance. III. New historical and chemical evidence on a wild Mediterranean dye-plant, *Daphne gnidium*', *Dyes in History and Archaeology* 16/17, pp. 5–9.
2. *ENPrESa: les Salviati 14e–16e; Entreprise, Négoce et Production en Europe (XIVe–XVIe siècles). Les compagnies Salviati.* Available at https://salviati.hypotheses.org/ (accessed 13 January 2021). The programme focuses on three main topics: account book keeping and monetary exchanges; trade networks and European territories; textile production and trade.
3. Cardon, D. (2007) *Natural Dyes: Sources, Tradition, Technology and Science*, London, pp. 177–9; Cardon, D. (2014) *Le Monde des teintures naturelles*, Paris, pp. 189–91.
4. Weight of sawwort bought, in Florentine pounds; 1 pound of Florence = 0.339 kg: Doursther, H. (1840), *Dictionnaire universel des poids et mesures anciens et modernes*, Brussels, p. 221.
5. Several different currencies were used simultaneously in Florence at the time, hence the need of such conversions: compare Goldthwaite, R.A. and Mandich, G. (1994) *Studi sulla moneta fiorentina*, Florence, Olschki. The florin *a oro larghi* was a gold currency.
6. All conversions from pounds to kilograms are approximate.
7. Doren, A. (1901) *Die Florentiner Wollentuchindustrie vom vierzehnten bis zum sechzenhten Jahrhundert*, Stuttgart, pp. 512–13.
8. Cardon 1995 (see note 1 above), pp. 62–3.
9. HBS Medici Family Collection, Baker Library Special Collections, Harvard Business School, MS551-1, 1502, f. 1r.
10. Doren 1901 (see note 7 above), p. 510.
11. Rosetti, G. (1969) *The Plictho of Gioanventura Rosetti* (Venice 1548) transl. S.M. Edelstein and H.C. Borghetty, Cambridge MA and London, p. 25.

12. Rebora, G. (1970) *Un manuale di tintoria del Quattrocento*, Milan, p. 10.
13. HBS Medici Family Collection (see note 9 above.)
14. Angelini, L.G., Bertoli, A., Rolandelli, S. and Pisteli, L. (2003) 'Agronomic potential of *Reseda luteola* L. as new crop for natural dyes in textiles production', *Industrial Crops and Products* 17, pp. 199–207.
15. Cardon, D. (1995/1998) 'La cuve de pastel: deux sources jusqu'à présent inédites / The woad-vat: two previously unpublished sources', in D. Cardon, H.E. Müllerott, B. Benjelloun, F. Brumont and M. Delmas, M. (eds), *Actes/Papers/Beiträge of the 2nd International Symposium 'Woad, Indigo and other Natural Dyes: Past, Present and Future' / 2ème congrès international 'Pastel, indigo et autres teintures naturelles: passé, présent, futur'*, Arnstadt, pp. 50–60.
16. Evans, A. (ed.) (1936) *Francesco Balducci Pegolotti: La pratica della mercatura*, Cambridge MA, p. 208.
17. Doren 1901 (see note 7 above), p. 512.
18. Archivio di Stato di Prato, Datini 312 *memoriale segnato* F, 1396; transcribed in Rebora 1970 (see note 12 above), pp. 9–10.
19. Hoshino, H. (1970) *L'Arte della Lana in Firenze nel Basso Medioevo*, Florence, pp. 238, 240, 243.
20. Scuola Normale Superiore di Pisa, Archivio Salviati, registers 397, 398 and 399, 1491–1493. See Houssaye Michienzi, I. (2015) 'Les milieux d'affaires florentins, le commerce des draps et les marchés Ottomans à la fin du XVe et au début du XVIe siècle', *Mélanges de l'École française de Rome – Moyen Âge* (online), No. 127-2. Available at: https://journals.openedition.org/mefrm/2753, https://doi.org/10.4000/mefrm.2753 (accessed 15 January 2019 and 13 January 2021).

Natural Yellow Dye Sources in Persian Carpets: A Review

Samaneh Sharif,[1] Paula Nabais[1] and Maria João Melo[*1]

Abstract Persian carpets are a precious product of art, culture and traditions of craftsmen. Colours have a substantial role in different patterns of the carpets, thus dyeing techniques have long been part of this unique industry in Persia. In this paper, we present a complete review of plants that were used as a source of natural yellow dyes in Persian carpets based on local and international references, including interviews with the few remaining dye masters in workshops located in the central parts of Iran. These results are compared with studies in which the yellow colours of Persian textiles were characterised by high-performance liquid chromatography with diode array detection (HPLC–DAD) and/or with the addition of mass spectrometric examination (HPLC–DAD–MS).

Introduction

Persian carpets are a precious product of art, culture and traditions of craftsmen and dyeing techniques have long been used to create their unique patterns. This paper reviews the plants that were used as a source of yellow dyes in these carpets and compares the results with studies that characterised the yellow colours of Persian textiles using high-performance liquid chromatography with diode array detection (HPLC–DAD) and with the addition of mass spectrometry (HPLC–DAD–MS). As well as local and international references, interviews with the few remaining dye masters in workshops in central Iran are included. Local sources of yellow dyes retained their importance and the identification of yellow dyes is useful for textile provenance studies.[1] Many plants in Iran could be regarded as natural sources of yellow dyes but there are few references that describe them together with their historical dyeing methods. Moreover, as the yellow dyes are more prone to fading, the actual yellow colours might appear different from the original colours.[2] The most common flavonoids used to dye yellow are flavonols and flavones.[3] The flavonol aglycones are characterised by a hydroxyl in position 3 of the pyran ring (C) and flavones by its absence (Fig. 1).

In flavones such as luteolin, in which there is no 3-hydroxyl group, the colorants create a stable yellow chromophore;[4] this is the main reason for

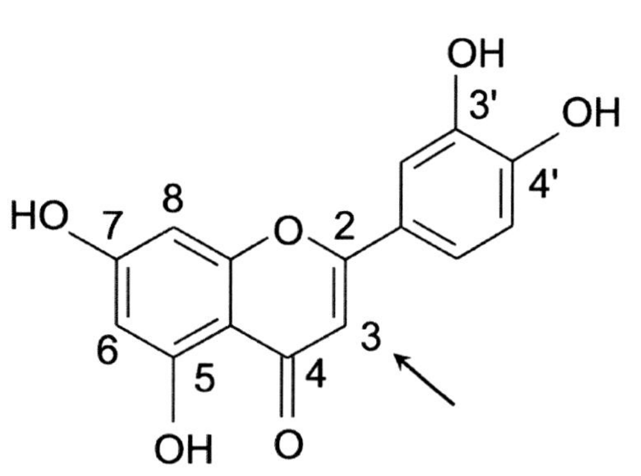

Figure 1 Aglycone structure present in flavonoid yellows; in position 3 an –OH group may also be present. In plants, flavonoids are found as glycosides.

Table 1 Names of plant sources of yellow dyes in Iran, geographic distribution, data on the mordants used, lightfastness and resulting colours

Botanical name	English name	Persian name and phonetics	Family	Distribution	Part of the plant used	Mordant	Colour on fibre	Lightfastness (1–8)
Delphinium semibarbatum	Larkspur	Zaban dar ghafa/*zæban.dær.qæfa*	*Ranunculaceae*	North-east of Iran	Flowers	Al	Yellow	3–4
						Cu	Bright pistachio green	5
						Cr	Opaque yellow	6.5–7
						Fe	Brownish grey	7–7.5
Eremostachys spp.	Desert rod	Sonbol-e-biabani/*:sɔ:nbɔ:l e bi:ɑ:bɑ:ni*	*Labiatae*	West of Iran	Leaves, stem	Not tested		
Pistacia vera	Pistachio	Pesteh/*:Pestɛ*	*Anacardiaceae*	North-east and centre of Iran	Hull	Al	Bright canary yellow	4–5
						Cu	Mustard green	7
						Cr	Mustard green	7–8
						Fe	Brownish grey	7
Punica granatum	Pomegranate	Anar/*ænɔr*	*Punicaceae*	Iran (most parts)	Peel of the fruit	Alkalis	Brownish red	Not tested
						Cu	Olive green	Not tested
						Sn	Yellow	Not tested
						Fe	Black	Not tested
Morus alba	White mulberry	Tūt-e- sefid/*tūt.e.sefid*	*Moraceae*	Iran (most parts)	Leaves	Al	Bright canary yellow	3–3.5
						Cu	Bright olive green	5.5–6
						Cr	Dark yellow lime	5.5–6
						Fe	Greyish brown	5.5–6
Prangos ferulacea	Prangos	Jashir/*dʒæshir*	*Umbelliferae*	Some parts of east, centre, south of Iran	Stem, leaves	Al	Bright canary yellow	3–4
						Cu	Bright olive green	5.5–6
						Cr	Sombre Hunyadi yellow	6.5–7
						Fe	Greyish brown	6.5–7
Quercus brantii	Oak tree	Dzaft/*dʒæft*	*Fagaceae*	East, south-east of Iran	Fruit/peel of the fruit	Al	Bright brown/Bright brown	3/4
						Cu	Brown/Brown	3–4/4
						Cr	Brown/Brown	4–5/5
						Fe	Grey/Dark grey	4–5/4
Reseda luteola	Weld	Isparak/*ispæræk*	*Resedaceae*	North-west, west, north-east of Iran	Blossom, stem	Al	Pure yellow	High
						Sn	Light yellow	Low
						Cu	Olive green	Low
						Cr	Olive green	Less than using alum
						Fe	Grey to brown	Low
Vitis vinifera	Grape	Angoor/ængur or *moʊ*	*Vitaceae*	Iran (most parts)	Leaves	Al	Cream yellow	4
						Cu	Pistachio green	4.5
						Cr	Dark yellow cream	5.5–6
						Fe	Brownish grey	5–6

the wide application of plants with this kind of chromophore, like weld, *Reseda luteola*, in dyeing. Flavonol aglycones, such as quercetin, kaempferol and rhamnetin, display a much higher sensitivity to light which could account for their inferior application in dyeing.[5]

Persian yellow natural dye sources

Our systematic research and interviews with dye masters indicate the use of the following plants as natural sources of yellow dyes in Persian carpets. The English and Persian common names are given (in brackets) after the scientific name. Table 1 indicates the names, geographical provenance, parts of the plant and the mordants used to dye as well as lightfastness data.[6] Lightfastness (right-hand column) was assessed according to the comprehensive study by Javidtash,[7] comparing the dyed textile samples to the eight ISO blue wool standard samples supplied by the Ciba-Geigy group: the higher the stated value (up to 8, the most stable to light) the more lightfast the dye-mordant combination described.[8]

Reseda luteola L. (weld; *isparak*)

The yellow flowers of this plant turn into small fruits with black seeds that can be planted. A biennial or perennial plant, its favourite location is the dark shaded areas near springs.[9] It has thick and strong branches, yellow or greenish-yellow flowers, leaves that are narrowly ovate or oblong-linear in shape and is 30–120 cm in height. Weld grows in the Ardebil, Lorestan and Khorasan regions of Iran[10] and is planted in early summer. The plants should be picked before the blossoms fall. All the aerial parts of this plant contain colorant materials, but the flowers are the main source.[11]

Armindo *et al.* studied a 16th-century Persian carpet that was preserved in the monastery of Santa Clara-a-Vella (Coimbra), before integrating the collections of the National Museum Machado de Castro (Museu Nacional Machado de Castro).[12] The results of their analysis of yellow dyes by HPLC–DAD demonstrated the existence of luteolin -7-*O*-glucoside as the main chromophore and lower amounts of luteolin, apigenin-7-*O*-glucoside and apigenin. Heitor *et al.* analysed samples from eight Indo-Persian carpets by HPLC–DAD–MS.[13] The presence of luteolin and its glucoside, luteolin-7-*O*-glucoside, in the expected proportions suggests *Reseda luteola* as a possible source, or a dyeing source with a similar profile, for all yellows. Hallet and Santos have also studied three Safavid Persian carpets from the palace of the dukes of Bragança in Guimarães, Portugal.[14] A luteolin-based dye (possibly weld) was the source of yellow dyes; it was mixed with indigo and madder to create green and orange respectively.

The main chromophores responsible for the yellow colour of weld extracts are luteolin and apigenin. Mordant dyeing processes on wool and silk provide a saturated yellow with very good lightfastness,[15] which is why it has been one of the primary colorants used for many centuries.

Delphinium semibarbatum Bien. ex Boiss. (yellow larkspur; *zaban dar ghafa*)

Delphinium semibarbatum (*D. zalil*) grows in deserts in the eastern and central parts of Iran; in Khorasan, Kashan and Yazd on plains and foothill meadows, where it reaches 30–80 cm in height and has pale yellow flowers.[16] According to Cardon, one of the larkspur's species with yellow flowers is called *isparak* in central Asia.[17] However, in Iran, botanists and researchers in the field of dyeing use *isparak* when referring to weld. The plant mentioned by Cardon, *D. semibarbatum*, is called '*Zaban-dar-qafa*' or '*Zaban-pas-qafa*';[18] *Zalil* is another common name.

In recent years in Iran, the use of *zalil* for dyeing has increased. The dye, which can be extracted from the flowers, leaves and stems of the plant, when applied on silk or wool with a mordant creates a bright and shiny shade of yellow with good stability.[19] When pomegranate peel or alum is used with a *zalil* dyeing, it also results in a bright and fast yellow.[20] The plant is rich in flavonols present as the 3-*O*-glycosides: quercetin, isorhamnetin and kaempferol. Only 5% of the glycosides are hydrolysed to aglycones during extraction from dyed silk.[21]

Kiumarsi *et al.* analysed the extraction process of *D. semibarbatum* for dyeing silk yarns:[22] their results

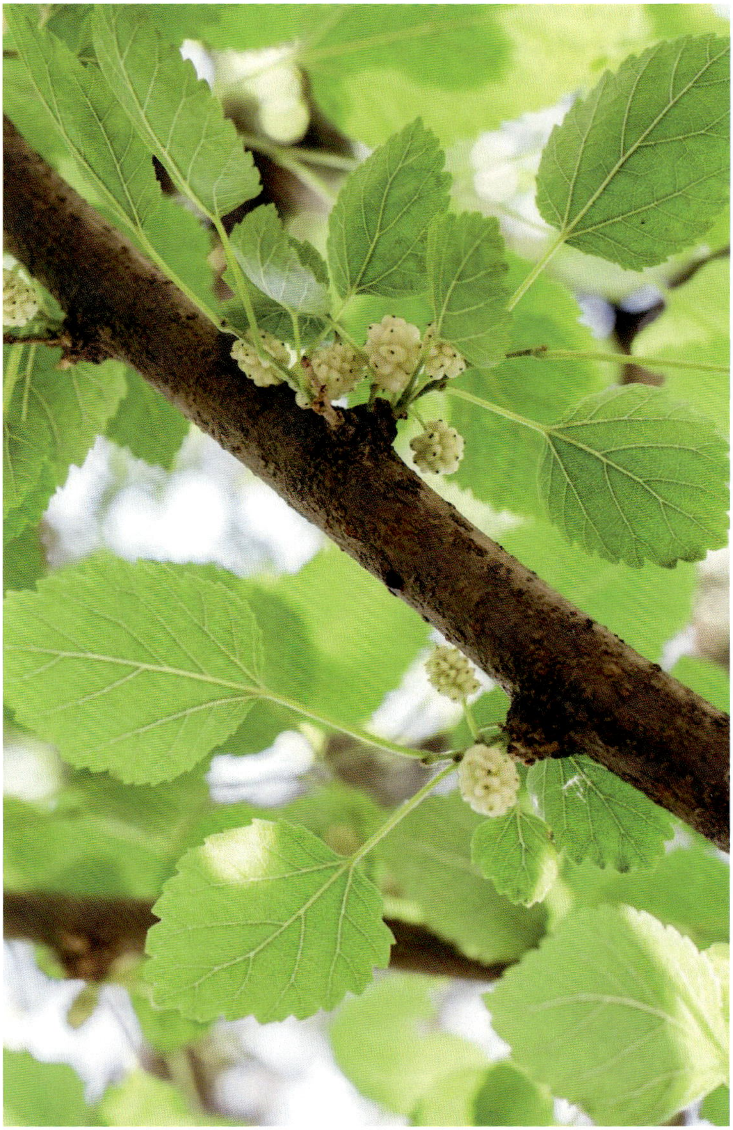

Figure 2 *Morus alba*. (Photo © Mahdyeh Sharif)

identified quercetin as the main yellow chromophore. Silk yarns dyed using the pre-mordanting technique possessed good lightfastness properties. Mouri *et al.* characterised the major components of *D. semibarbatum* and found the 3-*O*-glucosides of kaempferol, quercetin and isorhamnetin in similar amounts in the extracts.[23] Their work confirmed the existence of the glycosides in the dye rather than the aglycones. *Zalil* has good lightfastness – the reason it is widely used for yellow colours could be due to the low content of unsubstituted flavonols in the dye.

The use of larkspur in Persian textiles has been discussed by Shibayama *et al.*,[24] who analysed Persian Safavid velvets using HPLC–DAD. Most yellow samples were dyed with *D. semibarbatum*. The authors identified quercetin 3-*O*-hexoside, kaempferol 3-*O*-hexoside and isorhamnetin 3-*O*-hexoside as the main colorants of larkspur, in agreement with the results of Mouri *et al.*[25]

Prangos ferulacea Lindl. (prangos; *jashir*)

Prangos ferulacea is a perennial and gramineous plant with bright green leaves, a tall stem 80–200 cm in height and yellow blossoms. It is seen in the Markazi, Fars, Boyer-Ahmad, Charmahal, Isfahan, Lorestan, Hamedan and Kerman provinces and also in part of the Alborz mountains in Iran which enjoy cool and mild summers.[26] There are 15 different species of this plant in Iran, which are used as forage. Species such as *P. gaubae*, *P. calligonoides*, *P. cheilanthifolia*, *P. tuberculate* and *P. crossoptera* are exclusive to Iran. Other species are also found in Anatolia, Caucasus, and central Asia.[27] The colorant materials, found in its leaves and stems, produce bright yellow to olive shades with different mordants. Dyers use a combination of yellow from *jashir* and blue from indigo to create a green colour in carpets.[28] Mouri *et al.* studied different *Prangos* species and found flavonol 3-*O*-glycosides as the main yellow dyes;[29] 3-*O*-glucuronides of quercetin, isorhamnetin or rutin were the main components. Since the dried plant lacks free aglycones it can be used directly for dyeing.

Vitis vinifera L. (grape; *angoor*)[30]

The leaves of the grape, which contain the colorant material, are reported to have been used in Persia.[31] The dyeing process is similar to weld and produces yellow colours with alum on protein fibres.[32] However, the yellow colour produced with grape leaves on an alum mordant is not recommended due to its poor lightfastness.[33] Mouri *et al.* identified quercetin 3-*O*-glucuronide as well as some other flavonol glycosides as the main components of *V. vinifera*.[34] This accords with the results by Böhmer that confirmed quercetin as the main chromophore,[35] although Downey *et al.* identified both kaempferol and quercetin glucosides in Shiraz grape leaves.[36] Shibayama *et al.* found that some of the yellow pile samples from Safavid velvets were created from a combination of larkspur and another flavonoid dye containing luteolin, luteolin 7-*O*-glucoside and apigenin.[37] These could be different plant dyes similar to the leaves of *V. vinifera*.

Morus alba L. (white mulberry; *tut-e-sefid*)

Wild mulberry trees are scattered in the northern jungles but the cultivated trees are found in Kordestan, Isfahan, north of Tehran, Kermanshah and other areas of steppes or mountainous regions.[38] The oval fruit, which can be found in different sizes, consists of small white drupelets; the leaves are heart-shaped with small jags and contain the colorant material (Fig. 2). They also provide nourishment for silkworms. The leaves of this tree give a yellow dye that can be altered to orange, green or even brown by using acids in the dyeing process. The dyes are characterised by a morin core and are applied to produce a yellow colour for natural fibres.[39]

Eremostachys spp. (desert rod; *Sonbol-e-biabani*)

Desert rod is one of the plants used as a yellow colorant in the west of Iran. It grows in Isfahan, Nahavand, Lorestan, Kordestan and Charmahal. Different parts of the plant contain colorant material, but most of the dyes are found in the flowers which are orange, yellow or white.[40] There are 15 species of this perennial plant in Iran, five of which are exclusively found in Iran: *E. pulvinaris*, *E. codonocalyx*, *E. hyoscyamoides*, *E. azerbaijanica* and *E. adenantha*. Other species also grow in Anatolia, Afghanistan, Caucasus, Iraq, Pakistan, Turkmenistan and Central Asia.[41] Light yellow, dark yellow and yellow colours are obtained using aluminium, iron and chromium ions as mordants, respectively. All these colours are shown to have medium lightfastness.[42] We were unable to locate and collect this plant in Iran to confirm its scientific and common names.

Quercus brantii Lindl. (oak tree; *dzaft*)[43]

This oak species is a tall tree with lobed leaves; its blossoms have a greenish-yellow colour and the fruit is semi-oval, covered by fluff. A native tree that favours mild climates, it can be found in forests in Iran such as Lorestan, Kohgilouye, Kermanshah, Kordestan, Boushehr and near Shiraz.[44] The dye, extracted from the parenchyma of the oak,[45]

Table 2 Iranian plant sources of yellow dyes identified in the literature together with the extraction and analytical methods used for their characterisation

Botanical name	Yellow dyes*	Molecular structures	Authors and analytical methods	Extraction methods
Reseda luteola	**Luteolin 7-*O* glucoside; Luteolin** **Luteolin 3',7-*O*-diglucoside** Luteolin 4'-*O*-glucoside	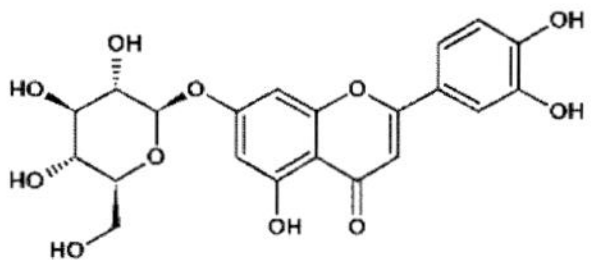	Wouters and Rosario-Chirinos 1992[46] HPLC–DAD	1 mg of yarn in 400 µL H_20/MeOH/37% HCI (1/1/2) for 10 min at 100 °C
			Pozzi 2011[47] SERS, with silver nanoparticles, λ_{exc} = 532 nm	6 mL MeOH and 200 µL 37% HCl at 65 °C for 60 min; the obtained solution was filtered and dried under a gentle stream of N_2
	Apigenin; Apigenin 7-*O* glucoside	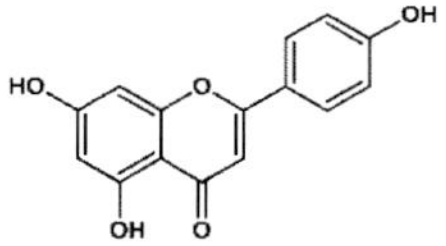	Moiteiro *et al.* 2008[48] HPLC–DAD (Dionex system) and HPLC–MS	*R. luteola* was extracted with 15 mL of MeOH:H_2O (8:2) by sonication at 25 °C for 10 min
	Kaempferol 3-*O*-glucoside-7-*O*-rhamnoside	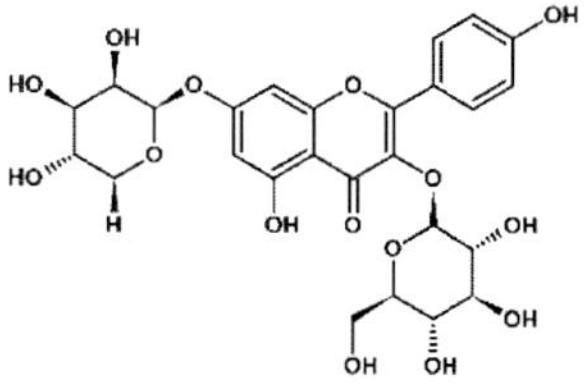	Shibayama *et al.* 2015[49] HPLC–DAD	0.1–1 mg with 100 µL of 0.001 M Na_2EDTA in deionised MeOH:H_2O (2/3) at 60–70 °C for 20 min. The sample was left for 1 hour in the solution; the extract was then evaporated. The residue was dissolved in 15 µL of 17.6% formic acid in deionised MeOH:H_2O (1/1)
Delphinium semibarbatum	**Quercetin 3-*O*-glucoside** **Kaempferol 3-*O*-glucoside**	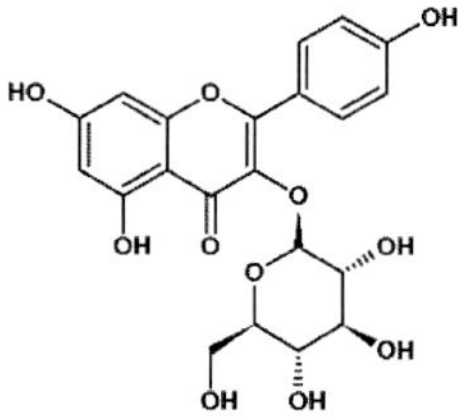	Mouri *et al.* 2014a[50] HPLC–DAD–MS	Extraction of raw material with MeOH:H_2O (1:1) at 65 °C for 1 hour. Dyes extracted from textiles with a 'soft method' (pyridine/water/1.0 M oxalic acid in water (95:95:10) at 100 °C for 15 min)
	Isorhamnetin 3-*O*-glucoside	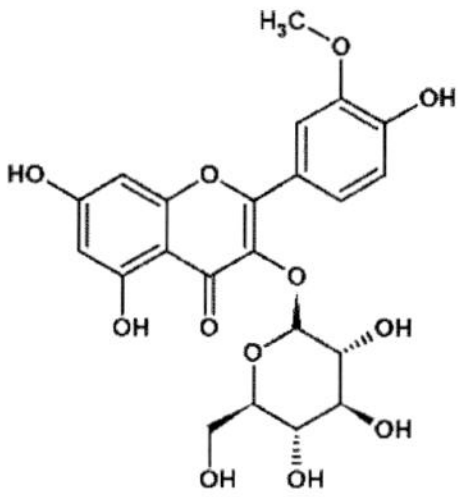	Shibayama *et al.* 2015[51] HPLC–DAD	0.1–1 mg with 100 µL of 0.001 M Na_2EDTA in deionised MeOH:H_2O (2/3) at 60-70 °C for 20 min. The sample was left for 1 hour in the solution; the extract was then evaporated. The residue was dissolved in 15 µL of 17.6% formic acid in deionised MeOH:H_2O (1/1)

Vitis vinifera	**Quercetin 3-*O*-glucorunide**	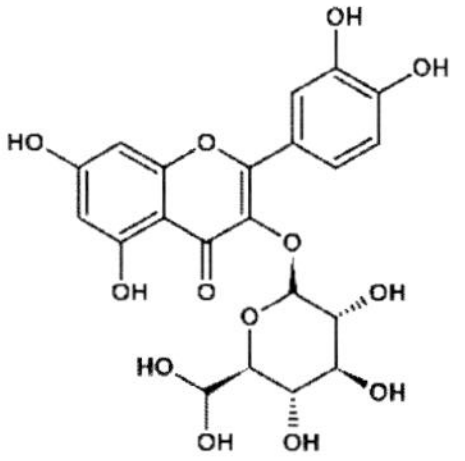	Mouri *et al.* 2014[52] HPLC–DAD–MS	Extraction with MeOH:H_2O (1:1) at 65 °C for 1 hour. Extraction of dyes with pyridine/water/1.0 M oxalic acid in water (95:95:10) at 100 °C for 15 min
Morus alba	**Astragalin** **Isoquercitrin** Moracin M Moracin C	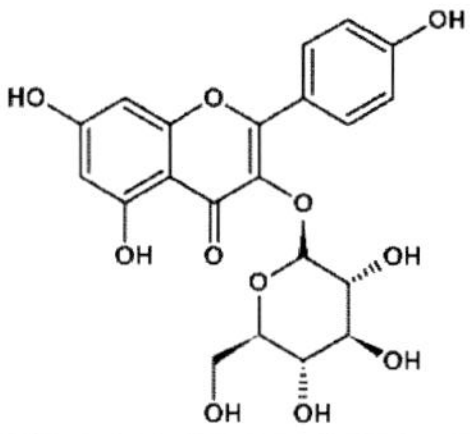	Yang *et al.* 2014[53] HR–ESI–MS and NMR. Preparative HPLC	The dried leaves were extracted twice (2 hours for each) with 95% EtOH under reflux and then the solvent was removed. The crude extract was suspended in water and extracted with ethyl acetate
Prangos Ferulacea	**Flavonol 3-*O*-glycosides** Quercetin 3-*O*-glucorunide Isorhamnetin/Rutin 3-*O*-glucorunide	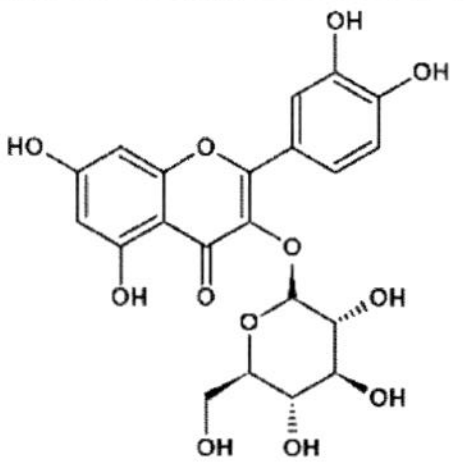	Mouri *et al.* 2014[54] HPLC–DAD–MS	Extraction with MeOH:H_2O (1:1) at 65 °C for 1 hour. Extraction of dyes with pyridine/water/1.0 M oxalic acid in water (95:95:10) at 100 °C for 15 min
Quercus brantii	**Quercetin** **Rutin** Flavonoid sulfate Flavon C and C-/*O* glycosides Apigenin, kaempferol, myricetin, naringenin	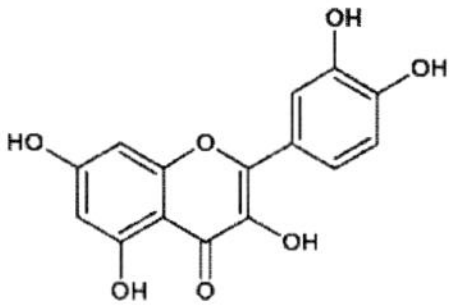	Noori *et al.* 2015[55] 2-dimensional paper chromatography. UV spec. for ID	Material was boiled for 2 min in 5 ml of 70% EtOH. The mixture was cooled and left to extract for 24 hours
			Çoruh *et al.* 2014[56] HPLC–DAD Other polyphenolic compounds are described	Extraction in methanol by using a rotating incubator at 30 °C and 175 rpm overnight. The initial extract was then filtered

* For each species, for the first name listed in this column, the molecular structure is represented in the third column. The main yellows, present in higher concentration, are in **bold**.

provides a yellow colour with aluminium ions as mordant. Other mordants are not applicable due to the lack of stability against light.[57] The dye belongs to the quercetin family and also contains tannin, which strengthens the colorant properties.[58]

Pistacia vera L. (pistachio; *pesteh*)

The pistachio tree grows in central and west Asia and the Mediterranean basin. Its height can reach up to 10 m in warm areas, but the trees that grow in mild climates sometimes resemble shrubs.[59] The hull (or husk) of the pistachio nut is used to dye yellow and pale brown; this is the outer fleshy layer of the pistachio that covers the hard shell containing the nut. Using different mordants can diversify the final colours. The soft red or yellow hull is removed during the harvest season and dried. It can be used as a natural source of yellow dye. According to a study by Kiumarsi and Parvinzadeh-Gashti, on wool yarns it produces light and dark shades of orange.[60] The authors found that the use of metal mordants with pistachio hull increases the dyeing ability properties and the colour fastness of the dyed yarns.

Punica granatum L. (pomegranate; *anar*)

The pomegranate grows in wide areas of Iran including the north, north-east, north-west, west and south. The tree blossoms in spring and fruits ripen in late summer to early autumn.[61] This plant contains a large amount of tannins. In the past, pomegranate peel was used in combination with sodium carbonate for cotton dyeing and when used with saffron it gives various colours. The peel of the fruit, which is removed and dried during autumn, contains brownish-yellow dyestuff. It is boiled in water and the remaining colour can be used after filtration. It can be applied without a mordant, but the yellowish-fawn colour is not durable.[62] Ancient civilisations in the Mediterranean were familiar with the use of pomegranate as a dyestuff. Applying mordants such as alum or iron would result in golden yellow or grey colours, respectively.

According to a study by Cardon, the colours from pomegranate were resistant to washing and light,[63] but Böhmer *et al.* reported an unsatisfactory lightfastness result.[64] Tiwari *et al.* found the fastness to light, washing and rubbing were very good in the case of wool but only just satisfactory when used on cotton fibres.[65] The pomegranate peel dye provided Mouri *et al.* with a yellow colour on cotton and brown on wool.[66] Middha *et al.* investigated the phenolic profile of *Punica granatum* fruit peel using high-performance liquid chromatography (HPLC).[67] They identified five different flavonoids, phenolic acids and their derivatives including quercetin, rutin, gallic acid, ellagic acid and punicalagin as a major ellagitannin.

Table 2 summarises the main chromophores of the dyes in the study and their molecular structures, as well as the analytical and extraction methods used.[68]

Conclusion

The lack of historical sources on traditional dyeing techniques in Iran meant that this knowledge was preserved by oral transmission only and remained unknown to most researchers. Furthermore, traditional dyers were unaware of the scientific names of the plants, which can lead to confusing information regarding the natural dye sources. As part of future research, we intend to comprehensively characterise the natural sources of Persian yellow dyes.[69] Following their validation by botanists, analytical studies will complete the process of identification and characterisation of dyes in historical textiles.

Editor's note and authors' addresses

*Author for correspondence: mjm@fct.unl.pt.
Paper received by 13 March 2018; revised version received 16 September 2018. This paper was presented as a poster at *Dyes in History and Archaeology* 36, Hampton Court, UK, 26–28 October 2017.

[1] Department of Conservation and Restoration and LAQV-REQUIMTE, NOVA School of Science and Technology, 2829-516 Monte da Caparica, Portugal

Notes and references

1. Heitor, M., Sousa, M., Melo, M.J., Hallett, J. and Oliveira, M.C. (2007) 'The colours of the carpets', in T.P. Pereira and J. Hallett (eds), *The Oriental Carpet in Portugal: Carpets and Paintings 15th to 18th Centuries*, Lisbon, Museu Nacional de Arte Antiga, pp. 1–9.
2. Ferreira, E.S., Hulme, A.N., McNab, H. and Quye, A. (2004) 'The natural constituents of historical textile dyes', *Chemical Society Reviews* 33, 6, pp. 329–36.
3. Cardon, D. (2007) *Natural Dyes: Sources, Tradition, Technology and Science*, London, pp. 167–484.
4. Zhang, X., Cardon, D., Cabrera, J.L. and Laursen, R. (2010) 'The role of glycosides in the light-stabilization of 3-hydroxyflavone (flavonol) dyes as revealed by HPLC', *Microchimica Acta* 169, 3–4, pp. 327–34.
5. Melo, M.J. (2009) 'History of natural dyes in the ancient Mediterranean world', in T. Bechtold and R. Mussak (eds), *Handbook of Natural Colorants*, Chichester, John Wiley & Sons Inc., pp. 3–20.
6. Mozaffarian, V. (1996) *A Dictionary of Iranian Plant Names: Latin, English, Persian*, Farhang Mo'aser, Tehran; Varzi, M. (1976) *The Art and Industry of Carpet in Iran: An Overview into History of Dyeing, Sketch and Texture*, Tehran, Raz Publications, pp. 141–84 (in Persian); Javidtash, I. (1996) *Dyeing Method with Plants and Stability Determination of 724 Natural Dyes*, Research Institute of Forests and Rangelands, Tehran, pp. 17–53 (in Persian).
7. Javidtash 1996 (see note 6 above).
8. The ISO blue wool standards are woollen fabrics dyed with eight different blue dyes of increasing fastness to light such that blue wool 1 is the least lightfast and blue wool 8 the most lightfast. For a given light exposure, blue wool 1 shows double the change seen in blue wool 2 and so on. Fading can be assessed by a colour measurement method or against a greyscale chart.
9. Cardon 2007 (see note 3 above), pp. 169–71.
10. Ghahreman, A. (2000) *Flore de l'Iran en couleurs naturelles*, Tehran, Institute des recherches des Forests et des paturage, Departement Botanique, vol. 20: 7, 16, 20, 23.
11. Jahanshahi-Afshar, V. (1996) *The Process and Methods of Dyeing Fibres with Natural Materials*, Tehran, pp. 38–107 (in Persian).
12. Armindo, E., Sousa, M., Melo, M.J. and Hallett, J.A. (2008) 'A Persian carpet's paradise garden: discovering historical and technical aspects through carpet conservation and restoration', in J. Bridgland (ed.), *ICOM Committee for Conservation, 15th Triennial Meeting, New Delhi, India, 22-26 September 2008*, 2 vols, New Delhi, vol. 2, pp. 960–66.
13. Heitor *et al.* 2007 (see note 1 above).
14. Hallett, J. and Santos, R. (2014) 'Interwoven knowledge: the understanding and conservation of three carpets', in A. Gerritsen and G. Riello (eds), *Writing Material Culture*, London, Bloomsbury, pp. 257–64.
15. Böhmer, H. with Enez, N., Karadağ, R. and Kwon, C. (2002) *Koekboya: Natural Dyes and Textiles. A Colour Journey from Turkey to India and Beyond*, transl. L.E. Fogelberg, Ganderkesee, REMHOB-Verlag, pp. 130–85.
16. Ghahreman 2000 (see note 10 above).
17. Cardon 2007 (see note 3 above), pp. 206–7.
18. Mozaffarian 1996 (see note 6 above).
19. Böhmer *et al.* 2002 (see note 15 above).
20. Wulff, H.E. (1966) *The Traditional Crafts of Persia*, Cambridge MA, MIT Press, pp. 189–93.
21. Cardon 2007 (see note 3 above), p. 207.
22. Kiumarsi, A., Parvinzadeh Gashti, M., Salehi, P. and Dayeni, M. (2017) 'Extraction of dyes from *Delphinium zalil* flowers and dyeing silk yarns', *Journal of the Textile Institute* 108, 1, pp. 66–70.
23. Mouri, C., Mozaffarian, V., Zhang, X. and Laursen, R. (2014) 'Characterization of flavonols in plants used for textile dyeing and the significance of flavonol conjugates', *Dyes and Pigments* 100, pp. 135–41.
24. Shibayama, N., Wypyski, M. and Gagliardi-Mangilli, E. (2015) 'Analysis of natural dyes and metal threads used in 16th–18th century Persian/Safavid and Indian/Mughal velvets by HPLC-PDA and SEM-EDS to investigate the system to differentiate velvets of these two cultures', *Heritage Science* 3, 1: 12; https://doi.org/10.1186/s40494-015-0037-2.
25. Mouri *et al.* 2014 (see note 23 above).
26. Ghahreman 2000 (see note 10 above).
27. Mozaffarian 1996 (see note 6 above).
28. Qaranjing, K., Karimshahi, K., Pirkohan, R., Arami, M., Kharati, H. and Kiumarsi, A. (2007) 'Using Taguchi method to optimize dyeing of wool yarns with jashir', *2nd National Seminar on the Hand-Knotted Carpet*, Tehran, pp. 123–31 (in Persian).
29. Mouri *et al.* 2014 (see note 23 above).
30. It is also called '/*:moʊ*/'.
31. Varzi 1976 (see note 6 above); 11. Jahanshahi-Afshar 1996 (see note 11 above); Wulff 1966 (see note 20 above).
32. Varzi 1976 (see note 6 above).
33. Böhmer *et al.* 2002 (see note 15 above).
34. Mouri *et al.* 2014 (see note 23 above).
35. Böhmer *et al.* 2002 (see note 15 above), p. 161.
36. Downey, M.O., Harvey, J.S. and Robinson, S.P. (2003) 'Synthesis of flavonols and expression of flavonol synthase genes in the developing grape berries of Shiraz and Chardonnay (*Vitis vinifera* L.)', *Australian Journal of Grape and Wine Research* 9, 2, pp. 110–21.
37. Shibayama *et al.* 2015 (see note 24 above).
38. Ghahreman 2000 (see note 10 above).

39. Varzi 1976 (see note 6 above).
40. Jahanshahi-Afshar 1996 (see note 11 above).
41. Mozaffarian 1996 (see note 6 above).
42. A research project by the Handicrafts Organization of Iran (1981), Tehran.
43. The Persian oak's fruit is called '/:b*ælūt*/' in Persian; but the part used for dyeing is the parenchyma, which is called '/:*dʒæft*/'.
44. Ghahreman 2000 (see note 10 above).
45. Parenchyma is the cellular tissue, typically soft and succulent, found chiefly in the softer parts of fruits.
46. Wouters, J. and Rosario-Chirinos, N. (1992) 'Dye analysis of pre-Columbian Peruvian textiles with high-performance liquid chromatography and diode-array detection', *Journal of the American Institute for Conservation* 31, 2, pp. 237–55.
47. Pozzi, F. (2011) 'Development of innovative analytical procedures for the identification of organic colorants of interest in art and archaeology' (PhD thesis, University of Milan), pp. 31–86.
48. Moiteiro, C., Gaspar, H., Rodrigues, A.I., Lopes, J.F. and Carnide, V. (2008) 'HPLC quantification of dye flavonoids in *Reseda luteola* L. from Portugal', *Journal of Separation Science* 31, 21, 3683–7.
49. Shibayama *et al.* 2015 (see note 24 above).
50. Mouri *et al.* 2014 (see note 23 above).
51. Shibayama *et al.* 2015 (see note 24 above).
52. Mouri *et al.* 2014 (see note 23 above).
53. Yang, Y., Yang, X., Xu, B., Zeng, G., Tan, J., He, X., Hu, Ch. and Zhou, Y. (2014) 'Chemical constituents of *Morus alba* L. and their inhibitory effect on 3T3-L1 preadipocyte proliferation and differentiation', *Fitoterapia* 98, pp. 222–7.
54. Mouri *et al.* 2014 (see note 23 above); Mouri, C., Aali, A., Zhang, X. and Laursen, R. (2014) 'Analysis of dyes in textiles from the Chehrabad salt mine in Iran', *Heritage Science* 2, 1: 20; https://doi.org/10.1186/s40494-014-0020-3.
55. Noori, M., Talebi, M. and Ahmadi, T. (2015) 'Comparative studies of leaf, gall and bark flavonoids in collected *Quercus brantii* Lindl. (Fagaceae) from Lorestan Province, Iran', *International Journal of Plant Research* 5, 2, pp. 42–9.
56. Çoruh, N., Nebigil, C. and Özgökce, F. (2014) 'Rapid and comprehensive separation for the phenolic constituents of *Quercus brantii* acorns by RP-HPLC-DAD', *Journal of Liquid Chromatography & Related Technologies* 37, 6, pp. 907–15.
57. Varzi 1976 (see note 6 above).
58. Jahanshahi-Afshar 1996 (see note 11 above).
59. Sabeti, H. (2008) *Forests, Trees and Shrubs of Iran*, Yazd, Yazd University Press, p. 257 (in Persian).
60. Kiumarsi, A. and Gashti, M.P. (2015) 'Pistachio hulls, a new source of fruit waste for wool dyeing', *Journal of Textile Science & Engineering* 5, 1; DOI: 10.4172/2165-8064.1000185.
61. Mirjalili, A. (2016) *Flora of Iran. Punicaceae family*, Research Institute of Forest and Rangelands, Tehran, 81, pp. 3–7 (in Persian).
62. Varzi 1976 (see note 6 above).
63. Cardon 2007 (see note 3 above), p. 482.
64. Böhmer *et al.* 2002 (see note 15 above), p. 156.
65. Tiwari, H.C., Singh, P., Mishra, P.K. and Srivastava, P. (2010) 'Evaluation of various techniques for extraction of natural colorants from pomegranate rind – ultrasound and enzyme assisted extraction', *Indian Journal of Fibre & Textile Research* 35 (September), pp. 272–6.
66. Mouri *et al.* 2014 (see note 23 above).
67. Middha, S.K., Usha, T. and Pande, V. (2013) 'HPLC evaluation of phenolic profile, nutritive content, and antioxidant capacity of extracts obtained from *Punica granatum* fruit peel', *Advances in Pharmacological Sciences* 2013: doi: 10.1155/2013/296236.
68. Mozaffarian 1996 (see note 6 above); Varzi 1976 (see note 6 above); Javidtash 1996 (see note 6 above).
69. Sharif, S., Nabais, P., Melo, M.J. and Oliveira, M.C. (2020) 'Traditional yellow dyes used in the 21st century in Central Iran: the knowledge of master dyers revealed by HPLC-DAD and UHPLC-HRMS/MS', *Molecules* 25, 908, pp. 1–16.